Series 7 Exam Prep

2024-2025

Ace the Series 7 Exam on Your First Try with No Effort | Test Questions, Detailed Answer Explanations & Insider Tips to Score a 98% Pass Rate

Amery Morgan

© **Copyright 2023 Amery Morgan - All rights reserved.**

The contents of this book may not be reproduced, duplicated, or transmitted without the direct written permission of the author or publisher.

Under no circumstances will the publisher or author be held liable for any damages, recovery, or financial loss due to the information contained in this book. Neither directly nor indirectly.

Legal Notice:
This book is protected by copyright. This book is for personal use only. You may not modify, distribute, sell, use, quote, or paraphrase any part or content of this book without the permission of the author or publisher.

Disclaimer Notice:
Please note that the information contained in this document is for educational purposes only. Every effort has been made to present accurate, current, reliable, and complete information. No warranties of any kind are stated or implied. The reader acknowledges that the author is not offering legal, financial, medical, or professional advice. The contents of this book have been taken from various sources. Please consult a licensed professional before attempting any of the techniques described in this book.

By reading this document, the reader agrees that under no circumstances will the author be liable for any direct or indirect loss arising from the use of the information contained in this document, including but not limited to - errors, omissions, or inaccuracies.

TABLE OF CONTENTS

INTRODUCTION: THE SERIES 7 EXAM 1
    CHAPTER 1: MASTERING BASIC SECURITY INVESTMENTS 9
    CHAPTER 2: INITIAL CUSTOMER OUTREACH 15
    CHAPTER 3: ACCOUNT OPENING AND MANAGEMENT 23

PART II: INFORMATION EXCHANGE AND CUSTOMER RELATIONSHIP 29
    CHAPTER 4: CONVERSATIONS ON INVESTMENT STRATEGIES 31
    CHAPTER 5: CONTINUING CUSTOMER RELATIONS 37
    CHAPTER 6: GOING TO MARKET, ORDERS AND TRADES 43

PART III: ADVANCED INVESTMENT PRODUCTS AND SECURITIES 49
    CHAPTER 7: UNDERWRITING SECURITIES: BRINGING NEW ISSUES TO MARKET 51
    CHAPTER 8: TYPES OF SECURITIES 57
    CHAPTER 9: ADVANCED AND SPECIALIZED INVESTMENT OPTIONS 65

PART IV: TRANSACTION HANDLING AND COMPLIANCE 73
    CHAPTER 10: TRANSACTION MANAGEMENT 75
    CHAPTER 11: COMPLIANCE, RULES, AND REGULATIONS 83

PART V: TAX CONSIDERATIONS AND RETIREMENT PLANNING 89
    CHAPTER 12: TAXES AND RETIREMENT 91

PART VI: FINAL PREPARATIONS AND EXAM TIPS 97
    CHAPTER 13: PORTFOLIO AND SECURITIES ANALYSIS 99
    CHAPTER 14: PRACTICE TESTS AND MOCK EXAMS 107
    CHAPTER 15: ANSWER KEYS AND EXPLANATIONS 129

CONCLUSION 141

INTRODUCTION

THE SERIES 7 EXAM

For anyone wishing to make their mark as a registered representative, the Series 7 test is one of those entry points into the banking business that can open up a wide range of prospects. The Financial Industry Regulatory Authority (FINRA) administers the exam as the starting point for any aspiring stockbroker in the United States.

Let's simplify things. The massive exam contains questions on various topics, including retirement plans, client accounts, equity and debt instruments, options, and taxation. Sounds difficult? That is true, so there. But the truth is, it's entirely feasible. It's difficult but possible with a passing grade of 72%. Thus, this exam has a way of leveling the playing field regardless of your level of experience in the banking industry or your desire to make a strategic career move.

Why, therefore, is the Series 7 exam so crucial? First, stockbrokers or registered representatives in the United States must comply with federal regulations. It's as easy as that: no Series 7, no trade. Beyond legal requirements, passing the Series 7 is like getting a badge of skill and credibility. It is a concrete endorsement of your knowledge and expertise in stock trading, sales, and all the rules governing these activities.

The exam format normally consists of several sections that evaluate both your knowledge and your application skills. In a little more than three hours of testing, there will be about 135 multiple-choice questions. It is a marathon, not a sprint, indeed. The Securities Industry Essentials (SIE) exam, the foundational test that assesses your fundamental knowledge of the securities industry, must also be passed before you may take the Series 7.

Most candidates dedicate months of diligent study to passing the Series 7 exam, which can require considerable preparation. But hey, this is your golden ticket to a thriving financial profession, deal-making, and perhaps realizing your Wall Street aspirations. Furthermore, you'll need to get extra licenses to carry out additional tasks, like selling different kinds of financial goods or giving investment advice, once you pass the Series 7. In other words, taking this exam begins your journey into the extremely lucrative yet hectic banking world.

The Series 7 is more than just a test. In the financial sector, it's a crucial milestone, a badge of honor, and, most importantly, the start of an exciting career path.

What to Expect on Exam Day: How to pass the exam
About the Series 7 examination, possessing knowledge of the anticipated circumstances on a significant day holds equal significance to comprehending straddles, calls, and bonds. Let us proceed directly to the subject matter at hand.

Firstly, it is advisable to engage in deep inhalation and exhalation. After extensive preparation spanning several months, the moment has arrived to assess and apply the acquired information through a formal examination. Upon arrival at the examination center, it is imperative to undergo a security check, thereby necessitating the possession of appropriate identification. Furthermore, it is imperative that any personal belongings, including mobile phones, be securely stored in the assigned locker. The Series 7 examination is quite rigorous in its emphasis on security standards.

The examination has a total of 135 items in the form of multiple-choice questions. The allotted time for the test is approximately 3 hours and 45 minutes, which includes a designated 30-minute break. Effective time management is crucial in this context, individuals. Every question is designed to evaluate your understanding and your capacity to apply that understanding in various financial contexts. The subjects encompassed under this domain span from stock securities and debt instruments to regulatory frameworks, consumer accounts, and taxation policies. To successfully pass, achieving a minimum score of 72% is important. Therefore, it is advised to ensure high confidence in one's responses before proceeding.

Now, how can one excel in this particular task? The approach encompasses a fusion of effective study techniques and adept test-taking methodologies. To effectively prepare for the exam, it is essential to develop a comprehensive study plan that encompasses a thorough exploration of the exam subject, supplemented by ample sample tests to facilitate familiarity with the question style and level of complexity. Many resources are readily available, encompassing textbooks, online courses, flashcards, and other materials. Select study aids based on your preferred learning style, such as being a read-write learner, a visual learner, or another type of learner.

Regarding test-taking tactics, addressing the questions that one finds most familiar and comfortable with is advisable since this approach can effectively enhance one's confidence. Next, go to the more challenging inquiries. When encountering a question that presents a significant difficulty level, it is advisable to refrain from excessively focusing. Please evaluate it and return to it at a later time. The efficient utilization of time is crucial, as it ensures that one can avoid preceding the opportunity to respond accurately to questions due to excessive time allocation towards challenging ones.

After responding to all the inquiries, utilize any remaining duration for a comprehensive assessment. It is advisable to thoroughly review and verify your responses, ensuring no questions were inadvertently omitted. Upon the expiration of the allotted time, the examination will be submitted, and the subsequent score report will be promptly sent.

Study Plan and Test-Taking Tips
You're determined to pass the Series 7 exam; you'll need a firm strategy to do so. This is a marathon, not a sprint; no one enters this beast of a test unprepared and slays it. Let's dissect the ideal study schedule and add some test-taking strategies.

Study Design:
1. Start studying at least three to four months before the exam date. This provides ample time to cover the wide array of Series 7 exam topics.
2. Resource Compilation: Assemble your study materials. This may consist of textbooks, online courses, or Series 7-specific applications. Choose your poison according to your learning approach; some people are bookworms, while others prefer interactive learning.
3. Create a weekly study schedule that includes the subjects to be covered, practice exams, and review sessions. Adhere to it like adhesive; consistency is essential.
4. **Bite-Size Learning:** Divide the topics into smaller, more manageable pieces. Focus on comprehending the 'why' behind each concept rather than simply memorization.
5. It's important to take practice tests and assessments to become familiar with the query format and identify your weak areas.
6. Devote time in the final month to reviewing and focusing on issue areas. You should also take full-length practice examinations under timed conditions at this stage.

Test-Taking Strategies
- **Time Management:** The Series 7 exam consists of 135 questions and lasts 3 hours and 45 minutes. That equates to approximately 1.6 minutes per query. Stick to this time frame, and don't get bogged down by any questions.
- **Rule of Elimination:** If you are uncertain about a question's answer, you should attempt to eliminate at least one or two incorrect choices. This dramatically improves your likelihood of guessing correctly.
- **First Instincts:** Your first instinct is frequently true. Don't overthink and confound yourself, which can lead to avoidable errors.

- **Flag for Review:** If you're truly stumped, mark the query for review and continue. You can return to it whenever time permits.
- **Remain Calm:** Anxiety can be a deal-breaker. Deep breaths can help reduce stress and enhance concentration.

You are setting yourself up for success by adhering to a meticulously planned study schedule and employing wise test-taking strategies.

How To Overcome Test Anxiety

Test anxiety is a real phenomenon, particularly when preparing for a career-defining exam like the Series 7. So, let's discuss how to keep your nerves in control and ace that test like a pro.

Preparation Is Crucial

First things first, being well-prepared can reduce anxiety significantly. The greater your level of preparation, the more confident you will feel walking into the exam room. This is why a comprehensive study schedule is essential. Observe your schedule, invest the necessary time, and ensure you cover every topic. Consider your practice exams to be the genuine thing. This will not only help you adjust to the pressure, but it will also give you a clearer idea of what to expect.

Mindfulness Methods

Meditation and deep breathing exercises can have miraculous effects on anxiety. Concentrating on your respiration and taking slow, deep breaths is key. This can reduce cardiac rate and enhance concentration. Try this in the days preceding your exam or even immediately before it begins.

Exercise

Engaging in physical activity is an effective way to relieve tension and reduce anxiety. Even a brisk 20-minute walk before an exam can help clear the mind. Endorphins, which are natural mood enhancers, are released during vigorous exercise.

Favorable Reinforcement

Your mental strategy must be solid. Continue to feed yourself affirmations of positivity. Tell yourself that you can do it because guess what? You do so. You have worked diligently and put in the necessary time to prepare for this.

Exam Day Customs:

Arrive early on the test day to acclimate yourself to the environment. Have a routine or ritual that helps you concentrate, such as eating a particular breakfast, listening to an energizing playlist, or donning lucky socks.

Take Time Off
If you experience test-related apprehension, do not hesitate to take brief mental breaks. Close your eyes, take several deep breaths, and then return to the queries. The clock is ticking, but a few seconds to regain composure can be decisive.

Maintain Perspective
Remember that this is only an exam. It is significant, but it is not the end of the universe. In the worst instance, you will take it again. You are much more than your test score.

Common Pitfalls
The Series 7 exam can feel like navigating a labyrinth; it is easy to become disoriented and even easier to fall into pitfalls. But have no fear! Knowing the prevalent pitfalls will allow you to avoid them gracefully. Here is a summary of what to look out for.

Negative effects of procrastination
Everyone has experienced delaying until the last minute to complete a task. Procrastination is your greatest enemy for a comprehensive examination such as the Series 7 exam. The exam covers a wide range of topics, from debt instruments to investment risk, and attempting to retain all of this information in a brief amount of time is a surefire way to fail.

Neglecting Weak Points
Everyone has both strengths and weaknesses. Some may be excellent with options but not with corporate securities. On test day, ignoring your vulnerabilities will come back to haunt you. Use practice tests to identify your weak areas and address them head-on.

Excessive confidence
Your performance on your practice exams is commendable. However, this should keep you from a false sense of security. Overconfidence can lead to less rigorous preparation, which is when unexpected events can cause problems.

Ignoring Test Techniques
Understanding the material is half the battle. You also need sound test-taking strategies to efficiently navigate the 135 questions. Not planning your time or employing strategies such as the process of elimination can result in the loss of valuable points.

Misreading Issues
In a stressful situation, it is simple to answer questions too quickly. Frequently, Series 7 contains queries that are intended to be difficult or confusing. Misreading can lead to selecting the wrong answer, even if you know the right one.

Absence of Review
You've answered all the queries; that's fantastic! But before you click the submit icon, carefully review your responses. Check for simple errors or queries that you may have missed by accident.

Anxiety and Tension
Last but not least, allowing stress to control you can impair your ability to think effectively and cloud your judgment. Find methods to manage stress before and on the day of the exam. Breathing deeply, taking brief mental pauses, and repeating positive affirmations can help keep stress at bay.

Insufficient Practice
There's no such thing as too many practice exams. The more you practice, the more familiar you will become with the question format, and the more proficient you will become at managing your time.

You are already ahead of the game if you are aware of these common pitfalls and actively strive to avoid them.

Time Management Strategies
Time management is the unsung savior of the Series 7 exam preparation process. This exam consists of 135 questions and lasts 3 hours and 45 minutes, so every minute counts. Let's explore some time management techniques to help you ace this task.

The Examination Phase
A thorough study plan that breaks down the extensive Series 7 curriculum into manageable sections is essential. Aim to cover topics well before the exam so that the final few weeks can be devoted to reviewing and practice exams.

When arranging your study sessions, set attainable goals. Overloading oneself is counterproductive. Concentrate on mastering one subject before moving on to the next.

Timed Practice Tests: Simulate the actual exam by timing yourself while taking practice tests. This provides a sense of urgency and allows you to practice your pacing.

The Day of the Examination
- Aim to arrive at the examination center at least 30 minutes beforehand. Utilize this time to unwind and mentally prepare for the upcoming task.
- Spend the first few minutes scanning the questions quickly. This will give you a sense of what lies ahead and enable you to allocate your time more efficiently.

Throughout the Exam
1. Break It Down: You are allotted approximately 1.6 minutes per question. Try to adhere as closely as feasible to this. Take your time, but don't linger excessively on a single query.
2. Easier First: Answer the queries you're most comfortable with first to earn points faster. This increases confidence and frees up time for more difficult problems.
3. Flag and Move: If a query stumps you, flag it and move on. The time wasted on a difficult query could have been better spent answering multiple simpler questions.
4. Elimination Game: When faced with difficult questions, use the process of elimination to narrow down your options. Even if you have to guess, your chances of being correct increase when you only have two options instead of four.
5. If time permits, return to the questions you marked for review and attempt them again. Even so, be aware of the time.
6. Remain Calm: Anxiety and distress are time-wasters. If you feel overwhelmed, close your eyes and take a few deep breaths. Reset and re-engage.

Time management is as important as understanding the distinction between a call option and a municipal bond. Mastering this skill will significantly improve your chances of passing the Series 7 exam.

Chapter 1

Mastering Basic Security Investments

Stocks and Bonds: An Overview
Let's get in-depth into the realm of stocks and bonds, two of the most fundamental types of assets you'll need to understand to pass the Series 7 exam. Understanding the subtleties of these instruments is absolutely necessary in finance because they serve as the industry standard.

The Stock Market Is a Game of Equity
When you buy a share of stock in a firm, you are purchasing a portion of ownership in that business. The ownership stake in the company entitles you to a proportional claim on the assets and earnings of the business. For this reason, stocks are also referred to as "equities"; when you own stocks, you have an equitable claim on the company's successes and failures.

Categories of Stocks
When discussing stocks, the common stock is the most commonly referred to type. This type of stock provides basic ownership rights, including voting rights at shareholder meetings and the possibility of receiving dividends. However, there is no guarantee for either.

Preferred Stocks are comparable to individuals invited to the exclusive VIP club of a company's stock offerings. Preferred stocks, on the other hand, often do not come with voting rights but offer a fixed dividend, and in the event of liquidation, preferred investors are given payment priority over common stockholders.

Capitalization of the Market
Market capitalization refers to the total value of a company's outstanding shares as determined by the market. This is often used to group companies into small-cap, mid-cap, or large-cap categories and can be a useful indicator of a company's level of risk and potential for growth.

Exchanges de Marchés

There are several different platforms on which stock can be exchanged, the most well-known of which are stock exchanges like the New York Stock Exchange (NYSE) and the NASDAQ. Each one is distinctive in its own right, thanks to the listing requirements and trading methods that are specific to it.

Over-the-counter, abbreviated OTC

Major stock exchanges do not trade all of the available equities. Some are traded over the counter through a group of broker-dealers' networks. These are typically more modest businesses, and as a result, they are subject to a greater degree of risk and fewer regulations.

The Borrowing and Investing Game Bonds

Unlike stocks, investing in bonds does not entail ownership of a business. By purchasing a bond, you lend money to the bond issuer, receive periodic interest payments, and repay the bond's face value upon maturity.

Various Forms of Bonds

- **Corporate Bonds:** Companies are the issuers of these types of bonds. When opposed to government bonds, they typically come with significantly higher interest rates to compensate for the increased risk level.
- **Municipal bonds** are debt obligations municipal, county, or state governments issued. These types of interest are typically excluded from taxation, making them appealing to investors with higher tax rates.
- **Treasury Bonds** are widely regarded as the most secure type of bond because they are issued by the United States government. They have interest rates that are lower than average, yet they are almost entirely risk-free.

The yield and the duration of it

The annual return on the bond expressed as a percentage of its present price is referred to as the yield. On the other hand, duration is a measurement that determines how long it will take for an investor to repay the bond's price from the entire cash flow it generates during its lifetime. Both are extremely important aspects that determine a bond's investment potential and hazards.

The Coupon Rate about the Yield to Maturity:

The yield to maturity (YTM) is the total return anticipated on a bond if it is held until it matures. This is in contrast to the coupon rate, which is the interest rate paid annually by the bond. The yield to maturity (YTM) calculation considers not only the bond's current market price but also its par value, coupon interest rate, and the number of years left until it matures.

Profiles of the Danger
Both stocks and bonds are associated with individual dangers and potential benefits. Investing in stocks typically results in bigger potential returns and higher levels of volatility. Bonds are generally considered risk-free investments because they give a stable source of income in the form of interest payments but have a reduced potential for producing big profits.

The act of diversifying
It is typically only suggested to invest some of one's money into stocks or bonds. A well-balanced investment strategy often involves holding some of each to reduce exposure to the risks associated with market volatility and to participate in the potential rewards associated with both asset classes.

Implications for Taxes
It is essential to have a solid understanding of the tax implications associated with equities and bonds. While the interest on certain bonds, such as municipal bonds, is exempt from taxation, the profits from the sale of equities are subject to the capital gains tax. Understanding how these investments may affect your current and future tax status can greatly impact the overall profits you receive.

Indicators of the Market
Monitoring market indicators such as the S&P 500, Dow Jones, and NASDAQ when trading equities is usual practice. When it comes to bonds, indexes such as the Bloomberg Barclays U.S. Aggregate Bond Index can offer extremely helpful insights into the tendencies of the market.

The yin and yang of investment alternatives are equities and fixed-income securities, respectively. A solid understanding of them is analogous to completing your fundamental training before venturing into the high-pressure realm of the Series 7 exam. As soon as you have a firm grasp of these principles, you will be well on your way to becoming an expert in the more difficult subject matter on the exam. You're not merely preparing for a test; rather, you're preparing for a future career in the financial industry. Consequently, you should fully comprehend these foundational components.

The Financial Markets: An Introduction
These marketplaces function much like hubs, where investors and traders congregate to buy and sell various assets. For anyone considering taking the Series 7 exam, having a solid understanding of the structure and function of various financial markets is like being familiar with your playing field before the championship game.

Contrasted with the secondary market is the primary market
The Primary and Secondary Markets will be separated, as this is the first order of business.

The primary market is the place where new securities are issued. In the primary market, corporations directly issue new stock and bonds to investors. This frequently takes place through transactions known as initial public offerings (IPOs) or direct placements. The issuing company receives all of the funds that are raised here.

After the securities have been issued, they will frequently be traded among investors in the secondary market. In contrast to the primary market, where investors provide their money directly to the company issuing the securities, investors in this market exchange their funds.

Exchanges for Stock
These are the environments in which buyers and sellers of stocks can conduct business. The New York Stock Exchange (NYSE) and the National Association of Securities Dealers (NASDAQ) are widely recognized in the United States. Other notable stock exchanges around the globe include the London Stock Exchange, the Hong Kong Stock Exchange, and the Tokyo Stock Exchange.

Order Varieties.
The following are some of the ways you can purchase and sell securities on various exchanges:

When you make a market order, you will buy or sell at the market's current price. If you place a limit order, specify the highest price you will pay or sell for. A stop order will only become active once the market reaches a specific price point.

Markets Available Without a Prescription
Over-the-counter (OTC) markets are decentralized markets that exist in addition to standard exchanges. OTC markets are marketplaces in which financial instruments such as stocks and bonds trade directly between two parties rather than on a central exchange.

The Foreign Exchange Market
Currency exchange takes place here. The foreign exchange market operates 24/7 (except on weekends) and responds quickly to geopolitical events.

Investing in Commodities
This is where the trade of tangible items, such as oil, gold, and agricultural products, takes place. Unlike stocks and bonds, commodities have a value that is not influenced by external factors and are more likely to be affected by supply and demand.

A Look at the Derivatives Market

On the market for derivatives, you won't actually be trading the asset itself; rather, you'll be trading contracts whose value is derived from the price of an underlying asset. Options and futures contracts are typical examples.

The Market for Fixed Income

This is the marketplace for issuing and trading debt instruments; another name for it is the bond market. Bonds can be issued by corporations, as well as by governments and municipalities.

Markets that are Bullish or Bearish

It is essential to have an understanding of the trends in the market:

- **Bull Market:** A bull market generally trends upward, and investor confidence is at its highest. The values of several securities are anticipated to go up.
- **Bear Market:** A market environment in which investor confidence is low and a decline in market prices is either currently taking place or is anticipated shortly.

There are two types of indicators

Market conditions in the United States are often assessed through indices such as the S&P 500, Dow Jones, and NASDAQ. These indices function much like thermometers, providing a brief summary of the overall state of the market.

Regulatory Authorities

Every player is responsible for familiarizing himself with the game's rules. The Securities and Exchange Commission, sometimes known as the SEC, serves as the primary regulatory organization in the United States. It is responsible for the regulation of the securities industry as well as the enforcement of federal securities laws. The Financial Industry Regulatory Authority (FINRA) is another crucial institution that governs member brokerage firms and exchange markets.

People who participate in markets

- Retail Investors are individuals like you and I who invest their own money.
- Institutional Investors Institutional investors are large investors such as pension funds, mutual funds, and insurance corporations.
- Speculators are investors who anticipate making a profit on short-term shifts in market prices.
- Arbitrageurs are people who aim to make a profit by taking advantage of pricing disparities that exist across many markets.

Understanding the financial markets is like having a solid grasp of the chessboard that the financial world uses. Every market category—whether for stocks, bonds, or derivatives—has its own distinct set of norms, developments, and participants. With all of this information under your belt, you are more prepared for the Series 7 exam and better prepared for a future career in the financial business. It will be a rough ride, but man, will it be worth it. So, pull up a chair and immerse yourself in this buzzing environment.

Chapter 2

Initial Customer Outreach

Let's delve into Chapter 2: Initial Customer Outreach, a fundamental component for anyone studying for the Series 7 exam and pursuing a career in finance. This stage is about making a good first impression, identifying customer requirements, and laying the groundwork for a productive relationship between the financial advisor and the client. So, how do you proceed? Let's disassemble it.

The Craft of Cold Calling
The infamous cold call is a requirement for many new financial advisors. It's more than just dialing a number; it's your first opportunity to pique the interest of potential customers. You should have a well-crafted pitch that goes right to the point and provides value immediately. Remember that people dislike feeling like they are being sold to, but they appreciate having their problems solved.

Consider Your Audience
It's important to investigate before communicating, whether a phone call or an email. Identify who you're speaking to. Are they Generation Y members who might be interested in ESG investing? Or a baby boomer concerned with retirement planning? Tailoring your approach based on their demographic and psychographic information can significantly impact.

The Ten-Second Pitch
When you finally get a foot in the door, either figuratively over the phone or in a face-to-face meeting, you have limited time to make your case. An elevator pitch can help you concisely describe your offer and why the prospect should care.

Client Needs Evaluation
The initial encounter isn't just about showcasing your capabilities but also about determining what the client requires. A requirements assessment is a requirement. You can use tools such as questionnaires and risk tolerance assessments to better understand your prospective client's investment objectives.

Building Faith
Let's face it: money is a touchy subject. Nobody will give their hard-earned money to someone they do not trust. Transparency is essential in these early phases. Clarify fees, potential returns, and associated hazards. Provide verifiable information, list your credentials, and do not overpromise.

Regulatory Conformities
When discussing Series 7, you can guarantee your bottom dollar that regulations will exist. Clients must be informed of the hazards associated with various investment options, and you may be required to provide certain disclosures. This violation can be detrimental to both your exam and your career.

Following up
Okay, the initial outreach was successful. What then? Immediate follow-up is required. This could be an email summarizing what was discussed, a next-step plan, or even a courtesy call. This serves two purposes: demonstrating professionalism and maintaining your name in front of the client.

Utilizing Engineering
In this digital era, we must recognize the importance of technology. CRM systems can be a godsend for keeping track of client interactions, requirements, and preferences. It facilitates not only organization but also customization of your approach.

The negatives
You will occasionally hear "no" -- perhaps more frequently than you'd like. It is an element of the game. The solution is to take it seriously and gain knowledge from it. You may need to revise your proposal or be better prepared for the client's objections. Each "no" brings us closer to a "yes."

The Game Plan
Initial consumer contact is not a one-time occurrence but the beginning of a long-term relationship. This is especially true in financial advice, where relationships can last for years or even decades. It is, therefore, less important to make a quick sale and more important to set the groundwork for ongoing financial guidance.

This concludes the overview of Initial Customer Outreach. This phase is your entry point into the lives of your prospective clients and an essential part of your Series 7 exam prep. Not only will this section help you pass the exam, but it will also set you up for a prosperous career.

Contacting Customers and Developing Marketing Materials
The Series 7 exam and your long-term career in financial services will require that you demonstrate that you can properly navigate this space. So, without further ado, let's get to the meat of the matter.

Contacting Customers Via the Following Methods

Let's face it: how you make contact is important. A call on the phone? Your email? A LinkedIn direct message that is skillfully crafted? The nature of the relationship might be influenced by the medium.

- Calling someone on the phone is an old but reliable method. It operates in real-time and affords users the chance to submit rapid input. However, time is crucial, and people want to avoid being bothered with a cold call while eating supper.
- Emails are fantastic for providing in-depth explanations and conducting follow-ups. What's the catch? The email you painstakingly crafted may be moved to the spam folder.
- Not only teenagers but also young adults and working adults can be found on social media. When it comes to reaching out to B2B clients, LinkedIn can be very helpful.

Putting Together the Message

How you say something is equally crucial as it is what you say. A message that has been carefully prepared may pique interest, impart knowledge, and motivate action.

Especially when dealing with cold outreach, it is important to get to the point as quickly as possible. Steer clear of jargon; it may make you sound intelligent, but it can turn off other people.

- The greeting "Dear Customer" isn't cut regarding personalization. You should use their names and tailor the message to each individual's specific concerns or situations.
- **CTA (Call To Action):** Direct the customer to the next step in the process; should they give you a call, click a link, or fill out a form? Make it simple and evident to everyone.

Documents Used in Marketing

Oh my, this is a significant one, particularly in a heavily regulated business like the financial services industry. Let's have a look at some of the most frequent kinds:

- It's old school, yet flyers and brochures can still be useful marketing tools for certain consumers.
- Websites and blogs are absolutely necessary in this day and age of digital technology. Points added for effective search engine optimization.
- Newsletters are excellent for maintaining engagement and education among prospective and existing customers.

It's all about compliance, compliance, and more compliance

It's important to remember that the financial industry's rules and regulations can be complicated. Any promotional content must meet the guidelines set forth by regulatory bodies like the SEC or FINRA.

Failing to comply with regulations can lead to significant fines and even disqualification from market participation.

Risk Warnings: There is always the possibility of losing money on an investment. Always include risk disclosures and avoid making statements not backed up by evidence.

Processes of Approval: Before marketing materials can be distributed to customers, compliance departments typically review and approve them.

Creating Successful Marketing Strategies
Don't just toss things against the wall to see what sticks. Research, planning, and measurement are the three pillars that support a successful marketing strategy.

Know who you are trying to contact regarding your target audience. Different demographics have varying requirements and respond to various methods in various ways.

The multi-channel strategy recognizes that different people have varying preferences. Some do well with emails, while others do better with social media.

Analytics are your best friend when it comes to metrics. Keep an eye on how effective your various kinds of outreach are, and make changes as needed.

Abilities in Communication
Although it may seem like stating the obvious, effective communication is much more than just relaying information from one person to another. It's about making connections, understanding, and growing your network.

To demonstrate active listening, try to speak less and listen more. Gain an understanding of the customer's needs, problems, and objections.

Always be sure to follow up with a feedback loop. Whether it be to dispel doubts, supply extra information, or simply express gratitude for their time, it is important to follow up.

Maintaining Your Knowledge
The world of finance is always evolving. Always be up to date on the latest market trends, laws, and changing requirements from customers. Webinars, newsletters, and industry publications are all useful tools for staying ahead of the competition.

Describing Investment Products to Customers

You've established that initial contact and created the basis; now it's time to get down to the "meat and potatoes," describing investment products to customers. This is a make-or-break phase, and acing it will not only be beneficial for your Series 7 but will also be essential for establishing trust and trustworthiness. Let's just jump right in.

Language Is Important Because

First, keep in mind that the person you are communicating with is not a Wall Street coworker; rather, you are conversing with a prospective customer who might not know the difference between a stock and a bond. It is not an example of dumbing down to use basic language rather than jargon; it is an example of being effective and courteous.

It is more effective to say, "This is a low-risk, long-term investment ideal for retirement planning," if you want to get your point across. This bond, which has a flattening yield curve and is rated AAA, is anticipated to perform better in a down market since the yield curve is flattening.

Analogies

Sometimes, using straightforward comparisons is the most effective approach to explaining a complicated idea to someone. Put purchasing bonds or stocks in the same category as buying a piece of a corporation.

These are the Big Four

Most investment portfolios include components from these four key asset classes: stocks, bonds, mutual funds, and exchange-traded funds (ETFs). The following is a breakdown of the different types:

- Explain that purchasing a company's stock is equivalent to purchasing ownership in the business. A greater potential for gain but also a bigger risk.
- Bonds can be loans to a corporation or government that accrue interest over their term.
- Smaller risk, but in most cases, it also means smaller rewards.
- Mutual funds are investment vehicles that allow investors to pool their money together to invest in a diversified portfolio of stocks, bonds, or other assets managed by financial professionals. They benefit investors looking to diversify their portfolios and lower their risk.
- ETFs, or exchange-traded funds, are investments comparable to mutual funds but traded in the same manner as stocks. They are, in general, less expensive and more adaptable.

Risk Accommodating Capacity and Time Perspective

When explaining investment options to clients, it's important to consider their risk tolerance and investment timeline. If they prefer a secure income with minimal risks, investing in Bondsay dividends may be the best option. On the other hand, if they are willing to take on additional risks for the potential of higher returns and have a longer investment horizon, growth stocks or aggressive mutual funds may be a better fit. It's crucial to align the explanation with the client's preferences.

Costs and Contributions
Ensure your customer is aware of any fees linked with the investment product, such as management fees, transaction fees, etc. This is more than merely considered good practice; it is typically required by various regulations.

Implications for Taxes
While other investment products may result in a tax burden for the investor, others come with tax advantages already attached. Two types of Individual Retirement Accounts (IRAs) for taxes are traditional and Roth. It is important to clarify this issue as it can make or break the agreement.

The Value of Having a Diversified Portfolio
Not only should you define the many investment products available, but you should also elaborate on the idea of a diversified portfolio. It's not wise to put all your eggs in one basket. Diversification is a strategy that can help control risk.

The Terms and Conditions
Oh, you're talking about the disclaimers. Every investment carries some risk, and you must ensure that the customer is aware of this fact. To reiterate, it's not simply the right thing to do; in many cases, it's also required by law.

To restate and briefly recap
It is always a good idea to recap the most important topics at the end of your presentation and then ask the customer if they have any questions. This acts as a recap and allows them to clarify any doubts they may have had previously.

Following the interview, sending a follow-up email summarizing the investment opportunities discussed might be of great assistance. It provides them with something to look back on and demonstrates your professionalism at the same time.

Continuous Instruction
The original description is merely the beginning of the process. It is essential to have regular communication with your customers regarding their investment goods so that they are aware of any changes in the market. Regular briefings, bulletins, or updates can be helpful in this regard.

An introduction to describing investment products to potential buyers is now complete. It is not enough to simply be knowledgeable in your field; you must also be able to communicate that knowledge in a way that is open, clear, and in tune with the customer's requirements.

Chapter 3

Account Opening and Management

Here, you should extend the "red carpet" to prospective clients rather than just saying "hi." This will help them feel welcome and like they belong in your financial world. Explore why this section is essential for passing Series 7 and your day-to-day as a financial advisor.

The First Paperwork
It's time for the less-than-glamorous paperwork now that you've wooed the client, and they're intrigued. Take advantage of this stage, though. The New Account Form serves as the guide for handling the financial destiny of your client, not only as a formality. It gathers crucial data, such as the client's financial position, investing goals, and risk tolerance. A long-lasting client connection can only be established by accurately filling out this form, which is essential for compliance.

Fit versus Fiduciary
A fiduciary standard and a suitability standard are distinct from one another. The appropriateness requirement requires advisors to suggest investment products appropriate for their client's financial needs. A fiduciary standard, on the other hand, requires advisors to put the client's interests ahead of their own and goes beyond that. Recognize whatever norm is expected of you and behave accordingly.

Customer knowledge (KYC)
This isn't simply a clever slogan; it's also required by law and considered best practice. Your ability to customize your advice and suggestions depends on how much you know your client's financial status, goals, and level of comfort with risk. Remember that KYC is a continuous process, and stay in touch!

Account Types
Give your client a breakdown of the many account types they can open, such as retirement, margin, and cash accounts. Each has benefits and drawbacks, legal requirements, and suitability for various investment goals. Understanding this inside and out makes it easier to match clients with the optimal account type for them and to respond to Series 7 queries.

Monitoring and Upkeep
Just opening the account is the initial action. Effective account management requires regular account reviews, rebalancing of portfolios, and updating of customer information. Clients should receive instructions on how to interpret account statements and any queries they may have should be answered simply and immediately.

Fixing issues
Last but not least, things can be challenging. Understand how to manage problems like account transfers, document typos, or client complaints. It's important to have problem-solving abilities because this is where you can wow clients the most.

To sum up, Account Opening and Management may appear to be one of the more uninteresting facets of financial counseling, yet it is essential. If you do this portion well, the relationship will be successful.

Informing Customers of Account Types and Providing Disclosures
This is more involved than just checking boxes on a form. Setting the stage now will help everything between you and your client go smoothly. So, let's talk about handling this important part of the account opening and management procedure and why having this information can help you ace your Series 7 exam.

Understanding Account Types
Ensuring the client is aware of their options is the first step. Each form of account, including cash accounts, margin accounts, and different IRA and Roth IRA types, has its own restrictions and advantages. A cash account, for instance, is straightforward but lacks the leverage that a margin account does. Retirement accounts offer tax advantages but also have withdrawal limitations.

You must provide thorough, understandable, and appropriate information for the client's needs and degree of understanding. Your goal is to direct them toward an account that matches their risk tolerance and financial objectives.

Information Disclosure
Greetings from the fine print world. Disclosures are essential to developing trust and not merely a legal duty. These documents detail everything, including the costs and dangers of various investment techniques. You must ensure the client understands these disclosures, not merely check the box to say they are okay. It's not only about avoiding legal problems; it's also about making sure the customer understands what they're signing up for, reducing the likelihood that they'll feel deceived in the future.

The "Why" of the "What"
It's a good idea to explain why each piece of information is important while making these disclosures. Does the client comprehend the ramifications of margin trading or the rationale behind some fees? Dissect it for them. This benefits you as well as helping to establish trust. The better judgments a client will make, the more informed they are, which will make your job easier.

Making It Continue
This is not an isolated incident. Updates might be required as laws change and the demands of the customer change. Good account management involves revisiting these themes frequently to ensure that the account type still matches the client's objectives and that they are informed of any new disclosures.

Regarding Season 7
Expect inquiries about various account kinds and the regulations that apply to them. Expect inquiries that will test your comprehension of how well you must enlighten your client. Don't just memorize the terminology; also comprehend their meanings and how they affect you as the advisor and client.

Updating Customer Information and Identifying Suspicious Activity
Let's move on to Updating Customer Information and Recognizing Suspicious Activity, another crucial aspect of Account Opening and Management. This may be a desk job, but it's much more than that. Understanding this is essential for your Series 7 and preserving your clients' assets and your company's credibility.

Refreshing Customer Data
Sure, life happens. People change professions, retire, get married, divorce, or experience other life-changing events that can greatly impact their finances. Because of this, it's essential to keep your clients' information current. Regular check-ins help you track any developments affecting your financial decisions. For instance, a client might need to take a more cautious approach to investing if they suddenly need to pay college tuition.

More than being a good counsel is required; regulatory agencies frequently demand that consumer information be updated regularly. Ask questions, take the initiative, and ensure you have the most up-to-date and complete picture of your client's financial life before accepting the information on the form.

Making Sense of Suspicious Activity
Let's now discuss the less amusing topics. Money laundering, insider trading, and financial fraud are all serious dangers that can have catastrophic consequences. Knowing how to recognize suspicious

conduct is important for more than simply compliance; it's also important for contributing responsibly to the financial ecosystem.

Anything from frequent, sizable withdrawals to a sudden interest in high-risk assets that don't fit a client's profile could be a warning sign. Other red flags include foreign transactions and frequent beneficiary changes. You must know these acts and report them according to your company's policies and any applicable legal requirements.

Documentation
Record everything, especially if you see anything fishy or whenever you update a client's information. Not only is it good practice, but you could also have to defend your choices to regulatory organizations.

Both of these topics are likely to come up in conversation. Know the legal requirements for updating customer information and the standard forms or steps to take. Be familiar with the warning signs and the proper procedures for reporting suspicious conduct.

In a nutshell, updating customer information and spotting suspicious activity are related. One involves adjusting to changes in your client's life and the marketplace; the other entails being watchful for dishonest or unlawful behavior.

Obtaining Customer Investment Profile Information
Let's obtain information about customer investment profiles, a crucial account opening and management component. Similar to the "getting to know you" phase of a new relationship, this section. Imagine it as the first in-depth discussion you have before becoming serious. And yes, it is equally crucial for the long-term sustainability of your advisory relationship as it is for your Series 7.

The Relevance
Let's first discuss why this is important. Your route map is the customer's investment profile, which describes their financial status, objectives, level of risk tolerance, and other preferences. You're not playing blindfolded darts; instead, you're aiming for a particular mark. You can give better investment advice if you have greater information. To ensure you offer investing options that meet their demands, there is a regulatory need to "Know Your Customer" (KYC).

Getting the Fundamentals
Gather the essential information first, including age, marital status, income, occupation, and tax status. These elements provide a foundational understanding of the client's financial situation and can aid in assessing the most advantageous investment plans.

Objectives and Risk Tolerance
You should now don your psychologist's hat. While some people enjoy roller coasters, others want to go skydiving. Investing follows the same rules. Recognize whether your client is risk-averse or is prepared to accept more volatility in exchange for bigger profits. Find out what their financial goals are. Are they saving for a round-the-world trip, a down payment on a home, or retirement? Knowing this enables you to provide recommendations that are more specific and useful.

Timeline
How much time is your client willing to commit? A different strategy will be needed for short-term goals like buying a car versus long-term ones like retiring.

Complexities and Legalities
Clients' special legal or tax circumstances, such as ongoing divorce proceedings, estate planning, or trusts, can occasionally impact investing decisions. To negotiate the challenges they add to the investment environment, be aware of them.

The Series 7 Viewpoint
Expect to be asked about the kind of data that should be gathered and why. Your exam will focus heavily on your knowledge of the KYC criteria and other regulatory obligations relating to consumer profile information.

As a result, when you're exploring the world of finance, consider gathering data on consumer investment profiles as your treasure map. You are more likely to discover the treasure—both in the shape of enduring, contented clients and passing that Series 7 exam—if it is drawn more precisely.

Obtaining Supervisory Approvals Required to Open Accounts
In the process of opening an account, this is when things really start to happen. It's more than simply a formality; consider it the quality control stage, ensuring everything is in order. Understanding this is essential for your day-to-day business and something you'll probably run against on the Series 7.

Justification for Supervisory Approval
To start, you could wonder, "Why do we need to take this step?" Supervisory approval, on the other hand, serves as a safety net for both the company and the client. It's an extra set of eyes ensuring that the account types, investing goals, and risk tolerance are appropriately and properly documented. This procedure ensures that all compliance and legal requirements are met.

The Command Structure
The paperwork often goes to your branch manager or another designated supervisor after you've gathered all the essential data from the client, filled out the necessary papers, and ensured that all

disclosures have been made and understood. This person is knowledgeable on both national and state laws, as well as particular business policies. They will examine everything to ensure that the suggested investment strategy fits the needs and profile of the customer.

Complete Checklist
The manager is confirming that you completed your homework. Have you accurately evaluated the client's risk appetite? Does the account type fit their financial objectives? Do any unique conditions, such as tax or legal matters? A supervisor could point out discrepancies or holes, request more details, or even ask you to change your recommendations.

Conformity, Conformity, Conformity
When a client's identification has been verified, and there are no suspicious circumstances, a supervisor will additionally verify for compliance with Anti-Money Laundering (AML) requirements at this step.

Series 7 and Real-world Consequences
You can encounter questions on the Series 7 exam about the approvals required to open various accounts or the approval process, including compliance considerations. Having a comprehensive understanding of this will benefit you not only in the exam but also in practical applications. The approval stamp gives you permission to proceed and confirms your thoroughness and diligence.

Supervisory approval is not just another administrative hurdle but a crucial component of the financial services sector's risk management and compliance culture. By mastering this phase, you'll prepare for the Series 7 exam and build the foundation for a career based on moral and effective practice.

PART II:
INFORMATION EXCHANGE AND CUSTOMER RELATIONSHIP

Chapter 4

Conversations on Investment Strategies

Let's examine Chapter 4: Discussions on Investment Techniques. Consider this chapter to be a private conversation you have with a client. It involves more than just demonstrating your familiarity with current market conditions or spouting trendy terms like "diversification" or "risk-adjusted returns." Here, you turn the financial goals of your customers into a detailed strategy. Additionally, you'll be tested on this in the Series 7 exam, so sit tight.

The Method
First and foremost, you should engage in these dialogues with a consultative rather than a directive style. Everyone dislikes a snob, right? Your objective is to direct the customer toward a plan that fits their objectives, level of risk tolerance, and time frame. This entails listening intently, posing meaningful questions, and being open about the advantages and disadvantages of various financial possibilities.

Discussing Objectives
The client's financial objectives, whether purchasing a home, setting aside money for college, or making retirement plans, should already be clear to you now. Match these goals with the right investment vehicles. Equities or long-term bonds may be a wise choice for someone who plans to retire in 30 years. However, more liquid and less risky assets can be the best choice for immediate objectives.

Risk versus Gain
Let's face it: talking about risk may be difficult. Some customers desire to reach the moon but might need more time to be ready for a possible crash landing. Dissect the risk-reward trade-offs between various tactics. A diversified stock portfolio, for instance, can have tremendous growth potential and market risk. On the other hand, a Treasury bond is safer but might offer more than the returns required to achieve its objectives.

Asset Management
Don't, to put it simply, put all your eggs in one basket. Discuss the significance of diversification, including geographical, internal, and cross-asset class diversification. Give examples of how a combination of stocks, bonds, and possibly alternative assets can provide a balance between risk and return.

The tactic aspect
After discussing strategy, let's move on to tactics. Discuss specific investment instruments that can be used to carry out the selected strategy. Your understanding of mutual funds, ETFs, specific stocks, bonds, and other instruments will be useful. Your task is to guide the customer through the complexity of each purchase's features, costs, and tax ramifications.

The Series 7 Relationship
These subjects will be covered in questions on your Series 7 exam. You must be able to adapt investment strategies to your client's needs, be familiar with the features of different investment products, and be conscious of ethical and legal issues when making investment recommendations.

Thus, Conversations on Investment Strategies can be considered the transition from theory to practice. You are more than simply a financial counselor; you are also a financial therapist, an educator, and a strategist. If you master this chapter, you won't just be getting ready for the Series 7; you'll also be positioning yourself for a job where you can actually improve people's financial situations.

Informing Customers about Investment Strategies, Risks, and Rewards
So, you've reached the stage where you must explain the intricacies of Investment Strategies, Risks, and Rewards to your clients. Consider this as a two-way discussion that is similar to dating. Ensure both sides are on the same page and speak frankly and honestly. This subject is crucial to your day-to-day work and a likely source of Series 7 exam questions. So let's get started!

Language Disparity
Keep in mind that not everyone talks "finance" first. Jargon may give the impression that you are intelligent, but it will only help if your client comprehends your words. So, make difficult terms simpler. Instead of stating, "Let's discuss your portfolio's asset allocation to maximize your risk-adjusted returns," consider adding, "Let's talk about the best mix of investments for you, considering how comfortable you are with risk."

Strategy Is Important
After the language is in order, you should discuss strategies. Do they value long-term development? Do they require a steady income? It could combine elements of both. Match the proper investment

vehicles to these objectives. A 10-year bond will probably not pique the interest of someone looking to purchase a home in five years. Your skills will be needed to match their needs with the appropriate investment possibilities.

Risk-Reward Game Playing
This is when things get a little complicated. All people desire high profits, but not everyone can tolerate high risks. Understanding both the potential drawbacks and benefits of any technique is crucial. Use examples from the real world or even make-believe situations to illustrate. Describe how stocks can be volatile yet have historically offered good returns. While generally safer, bonds may not be able to fight inflation. Although you are familiar with the procedure, your client might need to be.

Products and Alternatives
Options discussion follows strategy and risk discussion. Describe the different investment items that would be suitable for them. For risk-averse people, it might be a diversified mutual fund, while for those ready to take on greater risk, it might be a more focused ETF. Ensure they know any fees, tax repercussions, and other "small print" information that may affect their investment.

Never-ending Communication
After you've presented your plan, the work still needs to be completed. Markets and people's lives both undergo change. Regular check-ins are essential to validating or updating an existing strategy in light of changing conditions. Also, establish trust and increase your client's confidence in their investing decisions by having ongoing dialogues.

Takeaway from Series 7
You'll be tested on these topics during Series 7 to see how well you can explain them to clients. Therefore, it's essential to comprehend complexity and simplify it for both daily work and exams.

It takes careful balance to tell your consumers about Investment Strategies, Risks, and Rewards without being overbearing. Your job is similar to that of a translator, taking complicated financial jargon and turning it into clear, useful guidance.

Reviewing and Analyzing Customers' Investment Profiles
With a thorough examination of Reviewing and Analyzing Customers' Investment Profiles, the Conversations on Investment Strategies chapter would stay on track. Put your detective hat on to understand this. You're piecing together hints about your client's financial situation to solve the mystery of the ideal investment plan. Shall we begin? Rest assured, this will be included in your Series 7.

The Foundational Pieces

Let's first discuss the real components of an investing profile. You should consider the client's financial circumstances, investment goals, tax situation, and risk tolerance, among other important factors. These two things together will serve as the basis for whatever strategy you create.

The Economic Situation

Investigate the client's present assets, obligations, and cash flow. Ensure you know their income, debts, and any upcoming financial obligations. Are they buried with student loan debt? Or they may receive a substantial inheritance. You can determine what investing strategy is viable with this financial snapshot.

Objectives for Investments

Next, the objectives! Whether they wish to increase their wealth, make money, or possibly safeguard their cash, each objective necessitates a different approach. For example, growth can tip the scales in favor of shares or real estate, but capital preservation might involve lower-risk investments like bonds or money market accounts.

Tax Situation

Even though taxes aren't the most exciting issue, you still need to discuss them. It is crucial to know your client's tax rate and any prospective tax benefits or liabilities that may impact their investment. Ensure you are current on this because the Series 7 will require understanding how various investments are taxed.

Risk Acceptance

The psychological aspect, ah! Some people can't stand losing even a penny, while others are adrenaline addicts who want to risk it all. Quantitative analysis and a straightforward heart-to-heart discussion are frequently used to determine risk tolerance. In this case, it's important to emphasize that bigger gains also come with higher dangers.

The iterative method

Remember that an investment profile is flexible. People's financial situations alter significantly when they get married, have children, change careers, or go through other life transitions. It is crucial to regularly assess the investment profile to modify the strategy as necessary.

Real Life & Series 7

Understanding the subtleties of an investment profile is essential because the Series 7 exam will probably evaluate your ability to compile and interpret this data. Additionally, this is not just academic material; it has practical applicability. Every new customer or routine evaluation offers an opportunity to exercise these analytical skills.

Your guide will be the customer's investment profile. The better you interpret it, the more accurately and fully you can direct your client to their financial "X marks the spot." Let's don our detective hats to uncover those hints and solve the case.

Providing Required Disclosures Regarding Investment Products

Don't be fooled by the formal tone; think of this as the "legal stuff" part. This section is equally as important as the strategy discussion or examining your client's financial situation. This information not only may appear on the Series 7 exam but also enables you to maintain open communication with your clients and foster confidence. So let's get started!

What Matters in Disclosures

Imagine purchasing a car only to discover later that it consumes gas like there is no tomorrow. You may be rather irritated. The same is true with investments. Your clients have a right to be fully informed about their investments. This covers all aspects, good, terrible, and ugly, whether the potential for substantial rewards or the inherent risks.

What to Reveal

You'll be talking about a variety of subjects. First, fees and expenses. Be clear about the customer's costs, such as account maintenance and transaction fees. The risks associated with investments must be disclosed, including those unique to equities, bonds, mutual funds, and other financial instruments.

Remember the tax consequences. Some investments could appear fantastic at first appearance, but they may cause a customer to fall into a higher tax bracket or result in unfavorable capital gains tax consequences. It's important to be clear about everything up front because this could be a big deal-breaker for certain folks.

What to Reveal

No one reads the fine print, even though it is there for a reason, so there. It is your responsibility to organize this important knowledge into manageable chunks. In certain cases, this may entail presenting a condensed synopsis of the key points or reading through the disclosure documents themselves line by line. Make sure to be thorough while still being comprehensible. Truthfulness is paramount.

The importance of timing
Giving disclosures shouldn't just happen once, at the start of your partnership. It's time for an update on the disclosure front whenever you suggest a new investment strategy or product. It maintains a high level of trust and a little shock.

The Series 7 Viewpoint
The Series 7 test will likely include disclosures-related questions. They might inquire about the kinds of information that must be released or the circumstances under which disclosures must be made. Make sure you understand this subject thoroughly because it's essential for compliance and moral behavior.

Final Reflections
Giving required disclosures about investment products is like giving the financial world the dietary data. They may not have been the primary factor in picking up the box, but they are still important information. Make this a normal part of your interactions, like discussing your risk appetite or investment objectives.

Chapter 5

Continuing Customer Relations

We are about to begin Chapter 5: Continuing Customer Relations, so ensure you have your seatbelt on. This is the point in the relationship with your customers where you transition from the honeymoon period into a more permanent partnership. It is of the utmost importance to ensure the happiness of existing customers and retain new ones, compliance, and ethics. You will want to stay awake during this one since not only is it likely to be on the Series 7 exam, but it's also essential to your future profession.

Establishing Credibility and Relationships
First things first: trust and rapport between the parties is the foundation of any healthy, long-term relationship. The fact that your customers have put their faith in you enough to entrust you with their money is not to be taken lightly. It is critical to have open channels of communication at all times. Whether through regularly planned check-ins, newsletters, or simply being available to answer questions, the goal is to make your clients feel appreciated and in the loop about what's going on with your business.

Review of Your Accounts
Prepare to roll up your sleeves, for it's time for some review! Regular account reviews allow you to reevaluate your client's current financial condition, as well as their investing goals and their level of risk tolerance. It's similar to getting your automobile serviced; it ensures everything is operating correctly and allows you to make any necessary adjustments. Significant life changes such as marriage, work moves, or a fluctuating market are potential catalysts for the requirement of an evaluation.

Managing Customer Complaints and Comments
No one is flawless, and errors are possible in any situation. The customer may be dissatisfied with the performance of their portfolio, or they may have had a less-than-desirable experience with customer support. Regardless, dealing with concerns in a kind manner is a required aspect of the job. Listen, take

responsibility, and make amends. Additionally, it presents a potential for further development. Think of the feedback as unpaid suggestions on improving the quality of the service you provide.

Compliance with Regulations

Now that we have your attention let's talk about the rules. Every encounter with a consumer should be conducted in a manner that is compliant with the standards that govern the industry. In our conversation, we emphasized the significance of maintaining accurate records, disclosing all necessary information, and aligning recommendations with the client's financial objectives and risk tolerance. Failure to adhere to these guidelines could result in severe legal consequences and harm your professional reputation. It is crucial to take these matters seriously.

Providing Information to the Consumer

Even while most of your customers won't be as fascinated by the complexities of the financial market as you are, providing them with even a little information can go a long way. As their guide on their financial journey, it's important to keep your clients informed about the latest market trends, investment opportunities, and potential risks or rewards. Your responsibility is to equip them with accurate information so they can make informed decisions.

The Strategy for Getting Out

Indeed, there are situations when romantic partnerships are severed. How you cancel an account is important, regardless of whether the client wants to go a new route or whether the choice was made jointly. Make sure that everything goes off without a hitch. Make sure everything is in order with the documentation, and give the customer all the information they require to make the change. A dignified departure leaves the opportunity for a subsequent re-engagement intact.

Considerations for the Series 7 Exam

Thely test your understlikely anding of ongoing customer service. This may include everything from how to handle customer complaints to the requirements for maintaining regular communication. Not only is this information crucial for practical use, but it's also part of the exam. So, it's a win-win situation to learn it well.

It doesn't matter how savvy you are with money; you're playing the game properly if you can handle your client interactions. If you do, your customers will be grateful to you for approaching it as a continuous process and a relationship that requires regular maintenance. In addition, you won't have anything to worry about regarding the Series 7 requirements. Now you know everything there is to know about Chapter 5 in a nutshell. Now that you've got that out of the way get out there and rock those customer relations!

Communicating with Customers about Account Information

Regarding your reputation as a financial advisor, "Communicating with Customers about Account Information" is a topic that can make or break your credibility. Not only is this significant for establishing client trust and maintaining communication with them, but you should also anticipate seeing a question or two on the Series 7 exam about it. Let's get down to business here!

The Significance of Being Able to Communicate Clearly

Let's get the fundamentals out of the way first. The act of merely sending an occasional email or making a brief phone contact is insufficient to qualify as communication. The key is to establish a direct line of communication through which information may move back and forth without interruption. Your customer should always be aware of the condition of their account, and neither should you be concerned about their current living circumstances or changes in financial goals. Both of you should be aware of all relevant information.

What Should You Tell Them?

Now comes the part where things get serious. You need to keep your customer up to date on the following information:

- **Performance of the Account:** Customers are always interested in hearing how well their investments are doing. Maintain consistent communication with them regarding the status of the project and any potential variables that may be affecting it.
- **Transaction Updates:** Whenever you make a purchase, sale, or adjustment to their portfolio, make sure they are informed. After all, it is their financial responsibility.
- **Structure of Fees:** No one loves it when there are hidden fees. Be clear about what they are being charged for and why the fees are being assessed.
- **Alterations in the Market:** It is not just polite but also anticipated that you will give them advance notice whenever there is a significant event taking place in the markets that has the potential to affect their assets.
- **Implications for Taxes:** When it comes down to it, what matters most is the net return. Maintain communication with them regarding the potential impact of their investments on their tax situation.
- **Legal and Compliance Matters:** Inform the customer if there are new regulations that may have an effect on their account or the investing possibilities available to them. You are the one who is most knowledgeable in this area.

The Medium Is Important

It doesn't matter if you're using an email, a phone conversation, a video conference, or even a good old-fashioned letter to communicate with someone; how you do so is almost as essential as the content of what you say. Make it simple for your customers to communicate with you in the manner that most meet their needs. There must be regularly scheduled updates, such as reports issued quarterly. However, you should ensure you are available for those impromptu chats where you get into the intricacies.

The Practice of Active Hearing
Communication is a street that goes in both directions. Always make it a point to encourage your customers to communicate any issues, complaints, or simply general feedback regarding their account. Help them to feel at ease while they are expressing themselves. Not only does this make customers feel valued, but it also provides important insights into how to serve them more effectively.

Regulatory Prerequisites
Oh, you mean the small print? The regulations governing the financial business often stipulate that you must keep appropriate communication with your customers. This may include notifying them of noteworthy account activities, giving them regular statements, or even ensuring they receive prospectuses or other legally needed paperwork. Be sure to dot all the i's and cross all the t's.

Implications of the Series 7
Certainly, this subject will be asked on the Series 7 examination. You may be questioned regarding how to handle confidential information, the categories of information required to be released to customers, and the timing of disclosures.

Concluding Remarks:
It is non-negotiable to maintain effective communication with your customers regarding the details of their accounts. It is a sound business practice and a fundamental component of providing ethical financial advice. Give this facet of client interactions the priority it warrants. You'll put yourself in a position to achieve sustained success in this sector over the long run.

Processing Customer Requests and Retaining Documentation
One thing is certain when working in finance and dealing with people's money that they've worked hard for: you can only afford to be smart when processing requests or keeping records. This is the behind-the-scenes work that might need to be more exciting, but it is absolutely necessary for the efficient operation of the business and, you got it, for passing the Series 7 exam.

The Fundamentals of Taking Care of Customers' Requests
You have a client who has indicated they are interested in taking action, such as purchasing, selling, or modifying their existing investment strategy. Where do you want to start? Ensure you have a firm grasp of what they require of you. This typically entails several different interactions to ensure no

misconceptions that could result in any financial mishaps. After you understand the request, it is your job to carry it out promptly and precisely.

Having the Appropriate Documentation
If something needs to be documented, it is assumed not to have occurred in finance. Every action you perform, such as confirming a trade or transferring funds between accounts, must be backed up by the appropriate documentation. It's not enough to simply keep your office in order; you also have to comply with the regulations set forth by the industry. You will need to be familiar with the different kinds of forms required for the different types of transactions, the proper way to fill them out, and the location where they should be filed.

The Process of Verification and Authorization
Make sure you have the necessary authorization under your belt before attempting any major maneuvers. This could be a written authorization or a verbal confirmation that was recorded. Always remember these principles and remember that the regulations may sometimes define the type of authorization required. In addition, the consumer's identity must always be confirmed to prevent fraudulent activity. The outside world truly resembles a jungle.

In Support Of
Documentation, documentation, and even more documentation are required. Every encounter with a consumer ought to be backed up, particularly the ones that result in transactions. Ensure a paper trail (or a digital one) can prove each action taken. This can be accomplished through written confirmations, recorded phone conversations, or emails. This is necessary for resolving any issues that may arise and for any potential audits, whether they be conducted by regulators or by your company itself.

Updates in Real Time
Customers in today's fast-paced environment anticipate receiving real-time or nearly real-time updates regarding their accounts. If they have submitted a request, they will expect a confirmation as soon as the request has been dealt with. In many current systems, clients can track the status of their requests in real-time. If your company could be more tech-savvy. However, a fast email or phone call can go a long way toward satisfying your customers' needs.

The Maintenance of Documentation
The government mandates that banks and other financial institutions keep customer records for a while. This varies from jurisdiction to jurisdiction, and depending on the sort of record, it could be anywhere from many years to indefinitely. Not only is this vital for compliance, but it is also essential for addressing any possible disagreements or problems that might come up in the future. Always be

aware of the retention procedures required by your company and the jurisdiction in which you operate.

Things to Consider for the Series 7 Exam
Consider considering these responsibilities in your day-to-day work because you'll be doing much of it in your profession. It should come as no surprise that the Series 7 examination will put your knowledge to the test. You may be questioned about the protocols for carrying out customer orders, the need for documentation, or the guidelines for maintaining customer data. If you have a strong understanding of this topic, you will have an advantage both on the day of the exam and in your day-to-day work.

Assisting customers with their demands promptly and maintaining accurate records are more than administrative tasks. They are essential components of customer relations and have the potential to have a big influence on a customer's overall experience as well as their confidence in your company and its offerings. In addition, you have to ace them to comply with the regulations and succeed on the Series 7 exam.

Chapter 6

Going to Market, Orders and Trades

Let's continue to the next chapter, "Going to Market: Orders and Trades." This is the meat and potatoes of the situation, the big event, the phase in which all of the preparatory work you and your customer have done together begins to bear fruit. But what do you think? Additionally, the Series 7 exam significantly emphasizes this particular domain. Now that we have that out, let's discuss the information you need and why it's so important.

Various Forms of Orders

First, a solid grasp of the various kinds of orders is essential. There are a few different types of orders, including market, limit, and stop orders. Each one has perks and drawbacks, and understanding when to employ one over another might make all the difference in the cost of the implementation.

- Market orders allow for immediate Execution at the price currently available. Easy to understand and execute, but highly susceptible to price swings in the market.
- Buy or sell at a predetermined price or better with a limited order. More control over the price, but there is no certainty that it will be carried out.
- Stop orders become active once a certain price is reached, converting them into market orders. Helpful for preventing further losses or securing previously made gains.

The Execution, as well as the Settlement

It is not enough to just place the order; what matters is how it will be carried out and settled. While market orders are normally executed quickly, the timing of limit and stop orders might be more variable. Settlement typically occurs on the second business day following the trade date (T+2); however, you will need to be aware if there are any deviations due to the individual securities being traded or any other particular conditions.

Locations of Trading
Consider the New York Stock Exchange (NYSE), the National Association of Securities Dealers (NASDAQ), and even over-the-counter (OTC) markets. Suppose you are familiar with the guidelines and peculiarities of the various trading venues. In that case, you will be able to make more profitable trades.

Confirmations Regarding Trade
Oh, the piles of paper! Following the completion of a transaction, a confirmation must be forwarded to the customer outlining the finer points of the transaction, including the kind of trade, the amount transacted, the price, and any fees incurred. Maintaining an open and direct line of communication with a customer or client is not only necessary from a regulatory standpoint but also considered good business practice.

Keeping an Eye on Things and Making Changes
There is no "set it and forget it" kind of transaction regarding trades. As the market conditions shift, you may need to modify any open orders or even place new ones to accommodate the requirements of your customers or the opportunities presented by the market. Your experience, as well as your commitment to maintaining open lines of communication, are both essential at this stage.

The Perspective of the Series
On the Series 7 exam, you should anticipate being asked many questions about orders and transactions. This could involve anything from recognizing different kinds of orders to becoming familiar with the processes involved in trade settlement. There is also a possibility that you will be asked scenario-based questions in which you will be required to recommend a particular sort of transaction depending on the customer's requirements or the market's circumstances. Therefore, you must have a good foundation in this material.

When it comes to the world of finance, orders and trades are where the proverbial rubber hits the road. All the planning in the world will only help if you know how to properly get your product to market, no matter how much you talk about strategy. In addition, demonstrating mastery of this component is essential to pass the Series 7 exam.

Market Orders
Understanding market orders is vital for your day-to-day work and the significant Series 7 exam, and it is fine if you are a novice or an experienced trader. Therefore, let's go through the essential information you need to know.

What exactly is an "order in the market"?

A market order is a simple instruction to purchase or sell a security at its current market price. The key aspect to remember is the word "immediately." You express your desire to enter or exit the market without delay. This is the ideal method when prompt execution is necessary.

Pros:
- **Speed:** Market orders are promptly executed, which can be huge if the market is moving quickly and you want to catch a wave or bail out before a crash. Market orders are processed in real-time.
- **Certainty:** If you want to be absolutely certain that your order will be carried out, the market order will be your best buddy. There is no time to waste while waiting for the appropriate circumstances.

Cons:
- **Price Volatility**: You won't have any control over the purchase price, but your order will be fulfilled without a problem. You run the risk of buying at the market's high or selling when it's at its lowest point if the market is volatile.
- **Slippage:** Refers to the instance in which your order is carried out at a marginally different price from what you had anticipated. It most frequently occurs in markets that are either highly liquid or very sparsely traded.

These Are the Mechanics

When you put in a market order, your brokerage will submit it to the exchange or other marketplace with the highest probability of quickly executing the trade at a price acceptable to the trader. Your market order will be matched with the earlier of any existing limit orders for the stock you are interested in purchasing if there are any limit orders for the stock at the time.

The Management of Risk

Because of how market orders work, effective risk management is absolutely necessary. Because market orders are typically executed fast, you should always know how quickly your account balance can shift. This is something you should keep in mind at all times. When you have finished performing all essential risk evaluations, you should use market orders. A recipe for disaster is placing market orders without giving them any thought.

Exam Position for Series 7

If you're preparing for the Series 7 exam, it's important to understand market orders. It is crucial to know how they work, their advantages and disadvantages, and when to use them. You may be asked to select the most suitable order type for a particular scenario during the exam. Therefore, it's vital to have a complete understanding of market orders to prepare effectively for the exam.

Market orders are like that fast and furious friend who is always up for an adventure. They are wonderful for having a good time, but they may get you into trouble if you aren't careful. When playing the trade game, knowing when and how to employ them can make all the difference in the world. They are efficient and rapid but come with individual risks and potential payoffs. On the other hand, they are quite effective. And you can bet your bottom dollar that understanding them will give you an edge, both in the real world and on the Series 7 exam. You can be certain that gaining an understanding of them will provide you with an advantage.

Limit Orders

Now that we've covered that, let's talk about the other hot topic in town: limit orders. These individuals are the ones who think, who plan, and who take risks after careful consideration. Limit orders are used when you want to buy or sell at a certain price, as opposed to market orders, which allow you to enter and exit the market at breakneck speed. You can also use this information to your advantage on the Series 7 exam. So fasten your seatbelts!

In the End, What Exactly Is a Limit Order?

A limit order requests your broker to only execute a trade at a specific price or better. You can set this price beforehand. When buying, determine your maximum spending limit. When selling, decide on the minimum price you're willing to accept. Is it easy to understand?

Pros:
- You have complete control over the prices and all other terms. Your transaction will only go through if the market reaches or exceeds the specified price.
- Friendly to Your Wallet: Knowing the most you can spend is especially helpful when making purchases because it keeps your financial plans in check.

Cons:
- There are no guarantees, and the possibility exists that your trade will only go through if the market meets your pricing.
- The opportunity cost refers to the possibility of passing up other profitable trading opportunities as you wait for the optimal price.

Achieving Success with It

Your trade will typically be ordered by price, then by the time it was entered into the waiting pool when you use a limit order. Therefore, if the stock you're interested in does hit your limit, but there are previous orders at the same price, the orders placed earlier will be filled before yours. But you shouldn't worry about it. Most platforms make it possible to view the number of orders in front of you.

Comparison between Limit and Stop
You could also come across something known as a "stop-limit order." This is quite similar to a combination of limit and stop orders. As soon as a predetermined stop price is achieved, the stop-limit order is converted into a limit order. It is possible to obtain even greater control, but doing so also raises the possibility that your order will not be carried out.

For those that want to be in Series 7
In Series 7, you will encounter limit orders and must be familiar with all aspects of these orders. You won't just need to know what they are; you'll also need to know when to use them, their advantages, and their disadvantages. They may even throw you a curveball with a scenario that asks you to suggest a sort of order for a client that does not exist. Prepare to deal with it.

Concluding Remarks
Limiting orders is the way to go when you want more control over your trading price. They provide a strategic method for entering and exiting positions, but to successfully implement this strategy, patience is required, and risk is involved. When it comes to your price, will the market bite? Are you going to be put on "read"? This is just a normal element of the game.

It is not enough to have a working knowledge of limit orders to do well on tests or sound intelligent in front of your peers. When it comes to trading, it's all about making well-informed and purposeful judgments. Mastering the limit order will get you one step closer to achieving success on the Series 7 exam, which, let's be honest, will want to see that you know your stuff here.

Priority, Precedence, and Parity
Priority, precedence, and parity are the three concepts of the "three Ps." These are the ground rules for conducting business in the trading sector, which dictate how orders are executed once they are submitted to an exchange. Believe me when I say that even though they might sound like legalese, you should pay attention to them. This is especially true if you are preparing for the Series 7 exam. Let's take them apart, shall we?

First come, first served is the order of precedence here
At its most fundamental level, "priority" refers to the circumstance in which one customer's order takes precedence over another customer's order that is placed at the same price. It's the equivalent of shouting "shotgun" when you're about to get into an automobile, and it's used in trading. The

purpose of this is to encourage merchants to submit their orders promptly. The quicker you are, the higher up the line you'll be able to move.

Priority: The Price Is the Primary Consideration
The word "Precedence" is the key to everything. The cost of the item is given primary consideration by this guideline. Therefore, a buy order that is placed at a higher price has priority over a buy order that is placed at a lower price, and a sell order that is placed at a lower price has priority over a sell order that is placed at a higher price. Put another way, you will jump to the front of the queue if you are willing to pay a higher price (or settle for a lower price when selling). It's all about the numbers, really.

Equal Treatment: Caring is Sharing
Finally, but certainly not least, we have something called "parity." In the extremely unlikely event that numerous orders have the same price and time stamp, they are considered to be on equal footing, which means that their execution is proportional. Consider it in terms of dividing a pie: Each customer will receive 25 percent of the total shares if there are four identical orders. The market tells everyone, "Hey, you're all pretty special."

How They Cooperate With One Another
So, how do all of these ideas go together? In most cases, the most important thing to consider is precedence. Always give attention to the lowest possible price. After that, the priority system kicks in, which places orders in a queue at the same price based on the time they were received. Parity determines how the prizes should be distributed when there are no clear winners.

Regarding the Series 7 Examination
Now, other than wanting to be an informed trader, why should you care about all of this? There is a good chance that questions about these ideas will be included in the Series 7 exam. You will likely be asked questions that test your comprehension of how these principles carry out commands. On the exam, you might be presented with hypothetical scenarios and asked to select the sequence in which certain criteria would be satisfied first, second, and so on.

Priority, precedence, and parity are the trading equivalents of traffic lights; they direct the flow of orders and ensure everyone has an equal opportunity to succeed. Whether you are trading for yourself, working on behalf of clients, or studying for the Series 7 test, having a good grasp of fundamental principles is essential for understanding how the magic happens behind the scenes in the financial market.

PART III:
ADVANCED INVESTMENT PRODUCTS AND SECURITIES

Chapter 7

Underwriting Securities: Bringing New Issues to Market

Let's look at the world of underwriting securities, essentially the backstage pass for VIPs when bringing new issues to market. This information from Chapter 7 will be on your Series 7 exam, so you'll want to pay attention!

Then, What Does Underwriting Entail?
Imagine you are a cutting-edge technology startup that has just developed a revolutionary new idea. You need money to expand, but you will need more than your savings in your piggy bank. Underwriters are the people who step in at this point. These financial virtuosos collaborate with you to issue new stocks or bonds and then introduce those securities to the public market for the first time. They do a risk analysis, determine the pricing, and, in essence, serve as middlemen between the company and possible investors in the company.

The Primary or Leading Underwriter
This role, often held by an investment bank, is analogous to that of a quarterback on the underwriting team. The lead underwriter is responsible for a variety of tasks. They conduct research on the firm, delve deeply into its financials, and work toward establishing a price that is reasonable and enticing to potential investors. They are the ones that ensure that the financial paperwork is in tip-top shape and that everything is by the legal jargon, such as the regulations set forth by the SEC.

Different kinds of underwriting
The concept of firm commitment is comparable to the concept of "going steady" in finance. The underwriter makes a commitment to buy all the securities and takes on all the risks if they cannot sell them to the general public.

- This relationship style leans more toward "let's just see where this goes" than anything else. The underwriter makes every effort to sell the securities, but they need to guarantee that they will be successful.
- Investors will indicate to the underwriter the maximum amount they are willing to pay for the security, and the price will be determined based on the level of interest. It's like eBay, except for stock trading.
- Shelf registration enables a corporation to prepare many securities for sale, but the company can only sell such securities until the market is ready.

Syndicates and other forms of sales groups

The process of underwriting does not involve just one person. It is common practice for the lead underwriter to organize a syndicate, or a group consisting of several underwriters, to assist in risk distribution and to ensure that the securities are sold. After that, there are selling groups, which are often broker-dealers and are responsible for the actual act of selling securities to the general public.

This is the Roadshow

This is the opportunity for the organization to show off its capabilities. Consider it a cross between a rock concert and a TED Talk, as its purpose is to excite potential investors about the offering being made. The company's executives, together with the underwriters, went on a road trip to meet with large institutional investors, analysts, and anyone else who would be prepared to part with some cash in exchange for a piece of the pie.

In preparation for the Series 7 Exam

Pay attention because having a solid grasp of underwriting is necessary to pass the Series 7 exam. You will be asked questions about the many forms of underwriting, the associated duties and responsibilities, and possibly even some scenario-based questions that require you to apply the information you have learned. So give this chapter as much attention as you've given anything else you've ever studied. Okay, maybe not quite to that degree, but you understand.

When it comes to the realm of finance, underwriting can be compared to a matchmaking agency. It brings together businesses flush with cash and investors searching for the next great thing. Even though it is a difficult process with many moving parts, a healthy market must work properly. Suppose you can get the hang of this. In that case, you will be in high demand in finance and be well-prepared to ace the Series 7 exam and become a licensed financial professional.

Initial Public Offerings (IPOs)

The Initial Public Offering, an IPO, is like the glitzy debutante ball for companies just joining the stock market. It's a process that can be both exhilarating and nerve-wracking, packed with opportunity but also rife with the possibility of failure. Suppose you are going to take the Series 7 exam. In that case, you absolutely need to know everything there is to know about initial public offerings (IPOs). Therefore, let's go to work!

IPO: The Necessary Steps

The question is, what exactly is an IPO? Simply put, it is the first public offering of a company's stock. A firm is considered to remain private up until the time that it launches its initial public offering (IPO), at which point it will have a very small number of owners, most of whom will be early investors and workers. The initial public offering transforms the business into a public corporation open to investment from members of the general public, who are also referred to as retail investors.

Why Should We Go Public?

There are many different motivations for a company to go public. They may need funds for expansion, or they may want to get their debt under control. When a company goes public, it allows the initial private investors to cash in part of their shares for what should hopefully be a tidy profit.

What the Underwriter Is Responsible For

The underwriter acts similarly as a host for the initial public offering. The underwriter is typically an investment bank, and their responsibilities include analyzing the firm, determining the first stock price, and determining the number of shares to be issued. They not only assist the company in meeting regulatory standards but also ensure that no regulations are breached in the buildup to the initial public offering (IPO).

Pricing the initial public offering

This portion leans more toward art than it does science. The underwriter and the firm's executives need to agree on an initial price for the company's stock that accurately reflects its current value and is competitive enough to entice buyers. When determining the price, they will look at financial documents, current market conditions, and other criteria, such as the demand in the market. They will also consider something referred to as "underpricing," which is the practice of fixing the price a little lower to ensure a successful initial trading day.

This is the Roadshow

Before the initial public offering (IPO), the company will do a "roadshow," another name for a marketing tour. They will attend meetings with institutional investors, market experts, and occasionally even media members to generate interest. This is the opportunity for the company to demonstrate its business strategy, objectives for the future, and, most crucially, why it is a solid investment.

Prospectus of the Red Herring
In the time leading up to the initial public offering (IPO), the underwriters will distribute a "red herring" prospectus. This is a document that will contain information on the offering, but it will not include definitive information such as the share price. It is a regular element of the initial public offering (IPO) procedure, even though the name "red herring" gives the impression that something shady is happening.

The Angle of the Series 7
The topic of initial public offerings will certainly be included in your Series 7 exam. It is important to understand the role of underwriters, the necessary registration and disclosure requirements, and even the specific details of how the initial stock price is set. Additionally, you may be asked to guide hypothetical clients regarding the potential benefits and drawbacks of participating in IPOs.

Once the IPO had been completed
After the initial public offering (IPO), the company must comply with the ongoing reporting obligations established by regulatory organizations such as the SEC. This includes both quarterly and annual reports that are all available for the public and investors to review. Since the corporation is now open to the public, it is, of course, required to address the concerns of its shareholders.

A fascinating aspect of the world of finance, initial public offerings (IPOs) are known for their high stakes, dramatic nature, and the possibility of significant gains or losses. Understanding initial public offerings is about more than passing the test as you prepare for your Series 7. This knowledge from the real world can help you advance in your profession, whether you are counseling clients or considering making your own investments in firms that have recently gone public.

Seasoned Offerings
Now that we've covered the glittery world of initial public offerings (IPOs) let's move on to the equally essential but little less glamorous realm of seasoned offerings. Seasoned Offerings are when a company that is already public returns to the market for more funding. Initial Public Offerings (IPOs) are similar to the rookie season for firms entering the stock market. Seasoned Offerings are the sophomore, junior, and senior years. This information will be on the Series 7 exam, so pay attention!

What exactly is meant by "Seasoned Offerings"?
The sale of new or tightly held shares by a firm that has already made its initial public offering (also known as a "Secondary Offering") is referred to as a "Seasoned Offering." This company has been around the block, has public shares already in existence, and is already familiar with the reporting responsibilities that come along with being a public company.

Why Should I Choose Seasoned Offerings?

The reasons behind serving seasoned food can differ from person to person. The business must pay off some outstanding debt, raise money for an acquisition, or invest in brand-new capital projects. Seasoned offerings aim to raise capital for specific corporate goals, whereas IPOs generate capital for general business objectives and provide liquidity to current shareholders.

Different Categories of Seasoned Offerings

- A follow-on offering is issuing more shares later in the offering process. In most cases, the cash is given to the corporation without intermediaries, as explained earlier.
- In a secondary offering, existing shareholders, such as firm executives or early investors, offer their shares for sale to the general public to liquidate their holdings. These shareholders will receive the cash; the corporation will not.
- This allows a corporation to prepare a bundle of securities for sale but only sell them when appropriate, similar to having dinner prepared and ready to throw in the oven. Shelf offerings are similar to having dinner prepared and ready to throw in the oven.

The Process of Underwriting Once More!

Like an initial public offering (IPO), seasoned offers frequently involve an underwriter, normally an investment bank. The underwriter contributes to determining the share price, manages the sale, and, in some cases, may even buy the shares themselves to resell to the general public. Since there is already market data on the company's shares, the pricing strategy may differ from that of an IPO because it can help establish whether or not a price is fair.

The Current Regulatory Climate

Companies that perform seasoned offerings must submit the relevant forms and provide a prospectus, just as they would if they were doing an initial public offering (IPO). On the other hand, the process is typically more streamlined and expedient than that of an initial public offering (IPO) since these businesses are already open to the public and have been meeting continuing reporting requirements.

The Benefits and the Dangers

Seasoned issues can dilute existing shares, which means that the portion of the pie you own will decrease. This can be a cause for concern for existing investors. Nevertheless, dilution may be countered by a rise in share value if the newly raised capital is used to fund successful growth projects.

In preparation for the Series 7 Exam

Be familiar with the differences between initial public offers (IPOs) and seasoned offerings and the many kinds of seasoned offerings that can take place. You will be asked about the reasons behind seasoned offerings and the potential benefits and hazards to investors.

The Crux of the Matter

After making their debut in the public markets for the first time, businesses that want to continue growing their capital bases might consider making seasoned offerings. They are a significant factor in the functioning of the financial markets, even though they can be quite complicated and bring about new dynamics within the financial structure of a corporation.

Understanding seasoned offerings will not only help you ace the Series 7 questions, but it will also provide you with a deeper understanding of the market mechanics and company strategies, increasing your value to customers and potential employers.

Chapter 8

Types of Securities

In the Series 7 exam, one of the most important subjects you'll cover is "Types of Securities," so get ready to dive deep into this subject. This is the material you will be exchanging, recommending, and maybe even dreaming about (hopefully the relaxing kind of dreams and not the anxious kind). You must have a solid grasp of various types of securities. Let's get down to business here!

Equities, often known as stocks
Do we all understand? Buying a share of a company's stock means owning a portion of that business. Stocks are categorized into common and preferred. Common stockholders have voting rights, while preferred stockholders usually don't but have priority in receiving dividends. It's straightforward.

Fixed-income securities, most commonly known as bonds
In this scenario, you will lend the money you've worked so hard to earn to an issuer, such as a government or a corporation. In exchange, you will get regular interest payments and the promise that your principal will be returned when the investment matures. There are several distinct types of bonds, including treasury bonds, municipal bonds, and corporate bonds, amongst others. Each variety has its own particular qualities as well as fiscal repercussions.

AKA: "mutual funds"
Mutual funds can be a good option if you want to invest in a diverse range of financial markets. This is because multiple participants pool their capital to purchase various assets such as stocks and bonds. Since mutual funds are managed by professionals, you don't have to worry about the specifics. However, you may need to pay a fee for this service.

Exchange-traded funds, abbreviated as ETFs
These are like the mutual funds' cool and hip younger brother or sister. ETFs are another method of pooling money to create a diversified portfolio; however, unlike individual equities, ETFs are

exchanged on exchanges. You will have more leeway to purchase and sell at various points during the day as a result of this.

Choices available

If you buy options for an asset, you can buy or sell it at a specific price by a certain date, but you're not obligated to do so. Options have various uses, including hedging, generating income, or speculating. However, they can be complex and risky, so it's important to approach them with caution.

Real Estate Investment Trusts, or REITs for short

Do you want to invest in real estate but wait to own a building? You should invest in REITs. You can purchase shares in the trust managed by these companies, which own or finance real estate. They provide a source of income in the form of dividends and have the potential to increase in value.

Futures and other Speculative Investments

These are agreements to buy or sell a predetermined quantity of a commodity, such as gold, oil, or agricultural products, at a predetermined time. They are not for those who lack courage, but if done correctly, they can be profitable or act as a buffer against other types of investments.

Additional Securities

Other unusual investments, such as money market instruments, variable annuities, and even cryptocurrencies, are available. These are just some of the exotic fruits in the basket. Even though they might not be the primary emphasis of the Series 7, having a fundamental familiarity with them will only help you in the long run.

The Angle of the Series 7

Prepare yourself for questions that will go deeply into the features, benefits, and risks of the various types of securities. You may also be asked scenario-based questions in which you must suggest an appropriate investment for a made-up client. Be prepared to address the tax ramifications of the various types of securities and how those ramifications fit into the different investment plans.

Now that you've got that out of the way, let's quickly go through the thriving market for securities you'll need to be familiar with for the Series 7 exam. You must become familiar with these concepts early on in your career in finance because you will be using them frequently.

Corporate Ownership: Equity Securities

This is one of those essential concepts you'll need to grasp for the Series 7 exam, and it's also fundamental to your future job in finance, so you must study it now. Hold on to your seats!

Open Market Trading

To begin, most of the time, when discussing stocks, they typically refer to something called common stock. Common stock is comparable to purchasing a small portion of a business. You are granted voting rights (often one vote per share), which allow you to participate in the election of the board of directors and other important company decisions. Regarding dividends and assets, common investors are at the bottom of the priority list if the company declares bankruptcy. On the other hand, when a firm is doing well, they often benefit the most from it.

Stocks with Preferences

This is the most prestigious part of the equities securities market. Preferred investors are given preference over common stockholders when distributing dividends and assets if the firm goes bankrupt. On the other hand, they do not typically have the right to vote. It's similar to being a favorite visitor at a party; you receive all the privileges but don't get to choose the music that plays.

What do you mean by Class A, Class B, and Class?

Some businesses create distinct classes of stock to allocate voting power among their shareholders. The voting rights that come with Class A shares are typically more robust than those with Class B shares, which have the basic voting rights. By utilizing this structure, certain stakeholders can keep control while at the same time raising funds.

Payment of dividends

Let's talk about those incredibly delicious returns on investment. These are payments made to shareholders by a corporation, typically out of the corporation's earnings. Depending on the profits of the company and its dividend policy, common investors might get dividends, or they might not. However, preferred stockholders typically get a predetermined dividend, which brings this kind of ownership closer to the characteristics of a bond in some instances.

The practice of stock splitting and buybacks

These corporate acts have an immediate impact on the total number of outstanding shares and their value. A stock split results in an increase in the total number of shares while simultaneously lowering the price of each individual share. The converse occurs when a firm engages in stock buyback: it purchases its own shares, reducing the total number of outstanding shares and potentially increasing the stock price.

The Dilution Process and Efforts to Counteract It

Businesses can raise capital by issuing more shares; however, doing so reduces the ownership percentages of those already possessing shares. When corporations issue equity instruments, antidilution measures are sometimes included in those securities so that shareholders are protected against a decrease in the value of their ownership stake.

In preparation for the Series 7 Exam
You will need to understand the qualities that distinguish common stock from preferred stock, the workings of the various classes of stock, and the nature of dividends. You should be prepared to answer inquiries requiring you to compare these various sorts of stocks and advocate investing in them based on a variety of investor profiles and goals.

A Few Parting Thoughts
Most of a company's ownership comes from the purchase of equity securities. They provide potential investors with the opportunity to participate in a company's expansion and, in some cases, receive a percentage of the company's profits. However, they come with their own set of dangers and features, which is why it is so important to understand them for the Series 7 exam and your larger career in finance.

Debt Securities: Corporate and U.S. Government Loans
Now, let's shift gears and discuss the second major actor in the securities industry: debt securities. These are your bonds, notes, and other important financial documents. Debt securities are a key component of the Series 7 exam and your career in the financial industry. They are just as important as stocks. Are you prepared to dive in? Let's get going!

Bonds issued by companies
There are corporate bonds, which are essentially just IOUs issued by firms. When you buy a corporate bond, you're essentially loaning money to the company that issued it. In return, you receive interest payments at regular intervals and the full repayment of the bond's face value once it reaches maturity. These bonds can have terms that last from just a few months to as long as thirty years or more.

Why do corporations choose to make bond offerings? To raise funds for various purposes, including research and development and acquisitions. But keep in mind that, compared to, say, bonds issued by the United States Treasury, corporate bonds often come with higher yields to compensate investors for taking on more risk. Your credit score will be helpful to you in this situation. You can get a sense of the risk associated with a bond from rating agencies such as Moody's and S&P.

The United States Government's Bonds
Discuss things from the opposite end of the spectrum. Investing in bonds issued by the United States government is one of the safest investment options. These bonds include Treasury notes, Treasury

bills, and Treasury bonds. Due to the full confidence and faith in the United States government, you can be assured that your investment will be returned to you unless the country declares bankruptcy. The maturities of Treasury bonds are typically the longest in the market, reaching up to 30 years. Treasury notes typically have terms between two and ten years in length. Treasury notes, which have one year or less maturities, are the short-term investment choices.

Bonds from an Agency
Be sure to pay attention to government-sponsored agency bonds, such as those issued by Fannie Mae and Freddie Mac. Although they are not directly backed by the United States government, they are regarded as fairly safe and offer slightly higher yields than Treasuries.

In preparation for the Series 7 Exam
Prepare yourself for questions that compare corporate bonds to securities issued by the United States government. What are the potential dangers involved? What are the repercussions for the tax code? You can also be required to recommend appropriate debt instruments based on the characteristics of certain investors. Are they someone who avoids taking risks and prioritizes a secure income? It's possible that buying U.S. Treasuries is the best option. Are they open to taking on a bit more risk to get bigger returns? Bonds issued by corporations might be the answer.

Risk as well as Returns
The general rule in this scenario is that greater risk will result in a greater return. Corporate bonds typically have greater yields, but they also come with the risk associated with the corporation, which can vary greatly. Although the rates on U.S. government securities are lower, they carry a significantly lower level of risk. This is the trade-off that you and any future customers will have to consider.

The Role That Time Plays
Bonds with longer maturities typically provide greater yields but are also more susceptible to fluctuations in interest rates. Studying for the Series 7 exam requires a solid understanding of the relationship between changes in interest rates and the prices and yields of bonds.

Debt securities are an alternative to equities that entails a lower level of risk; nonetheless, they are more complicated and have more complexities, which call for an in-depth understanding. You will need to have a thorough understanding of them to effectively recommend a diversified bond portfolio or even to hedge against the volatile nature of the market.

Municipal Bonds: Local Government Securities
Let's look at municipal bonds, which are an important part of the overall landscape of the debt securities industry. Even while you may hear less about them than you do about their corporate or

federal equivalents, municipal bonds are something you should definitely pay attention to, especially for the Series 7 test. So, let's get this show on the road!

What exactly are local government bonds?

Municipal bonds are issued when local governments such as towns or states must raise money for public projects such as building schools, roadways, or sewage systems. Municipal bonds can be purchased at a fixed interest rate. When financing their day-to-day operations or particular projects, these bonds are a vital instrument for the companies in question.

Two Primary Categories: Bonds of General Obligation and Revenue Also Available
To begin, there are General Obligation (GO) bonds, secured not only by the issuer's full faith and credit but also by the issuer's ability to tax. This indicates that the local government can collect taxes to repay the bond. When you purchase a GO bond, you are placing a wager on the overall financial health of the corporation issuing the bond.

Then, there is a type of bond called a Revenue Bond, which is slightly more detailed. These are paid back from the money that comes in as a result of the particular project the bond is supporting. Imagine roads with tolls or a brand-new sports stadium. Unfortunately, the project needs to create a sufficient amount of revenue. In that case, you are still looking for a way to get through.

Advantages for Taxpayers
The interest income generated from municipal bonds is typically exempt from federal taxes, and in certain cases, state and local taxes are among the municipal bond market's most attractive features. Because of this, they are an especially attractive option for investors who are in higher tax rates. You will need a solid understanding of the tax-equivalent yield to pass the Series 7 exam and move forward in your career as a financial advisor.

Factors of danger
Municipal bonds are not risk-free investments, even though they are often considered less hazardous than corporate bonds. The importance of the issuer's creditworthiness cannot be overstated. Default by local governments is possible and does occur, albeit not very frequently. Additionally, because many municipal bonds have long maturities, they are susceptible to fluctuations in interest rates.

In preparation for the Series 7 Exam
You will need to be able to analyze the quality and risk of municipal bonds, distinguish between GO bonds and Revenue bonds, and comprehend the tax consequences for investors to succeed in this field. You can be asked to suggest an appropriate form of bond for an investor depending on their level of risk tolerance, their current tax status, and the income requirements they have.

Considerations Regarding Liquidity
Municipal bonds may have a lower liquidity than other types of assets. There is a secondary market, but it is less busy than the markets for corporate bonds or Treasury bonds. This could result in a bigger gap between the bid and the ask price or greater price volatility.

Bringing It All Together
You must have a solid understanding of municipal bonds for both the Series 7 exam and your future employment in financial services. You will need to be familiar with how they differ from other forms of debt securities, the risks involved, and the methods by which you can determine the tax-equivalent yield for your customers. They provide a one-of-a-kind balance of danger and reward that may be fashioned in such a way as to cater to the requirements of particular investors, particularly those who are searching for tax advantages.

Consequently, remember the municipal bonds you need to purchase to pass the Series 7 exam. Understanding them can be the key to developing a well-rounded investment strategy that takes advantage of the unique flavor they provide to the world of debt instruments.

Chapter 9

Advanced and Specialized Investment Options

Let's move on to Chapter 9, exploring the more intricate, niche world of investing alternatives. You should pay particular attention when preparing for the Series 7 exam because questions on these subjects can become quite complex. Snug up!

Options
Investing in options can be a thrilling experience, similar to riding a roller coaster. Options are financial products that grant you the right to buy or sell an asset at a specific price within a set time frame. However, options are not for the faint of heart, as they come with significant risk and the potential for significant returns. To invest wisely, it is essential to have a thorough understanding of premiums, strike prices, expiration dates, and the difference between call and put options (buying versus selling).

Investment funds
Mutual funds play a crucial role in many financial portfolios, even though they are less complex than options. They pool funds from various investors to purchase a diversified portfolio of stocks, bonds, or other securities. Familiarizing oneself with the types of mutual funds, fee structures, and pros and cons of other investment options is essential for the Series 7 exam.

Exchange-traded funds (ETFs)
ETFs are a hybrid, giving the diversification of mutual funds with the liquidity of individual equities. They are similar to mutual funds but traded on stock exchanges. Understand their differences from mutual funds, particularly in costs and tax effects.

Real estate investment trusts, or REITs
Have customers who are interested in real estate but are hesitant to purchase a property directly? Real estate investment trusts (REITs) provide a more liquid alternative to investing in real estate. Equity REITs, mortgage REITs, or a combination of both are possible.

Hedging Funds
You can invest in these high-risk, high-reward pools. They typically have high costs and are only accessible to accredited investors. To obtain significant profits, hedge funds employ various methods, including short selling and leverage. However, enormous promise also entails big risks.

Annuities
Pay attention to annuities, agreements that guarantee a series of payments at regular intervals. You'll need to be familiar with various types, including fixed, variable, and indexed annuities, as well as their benefits and drawbacks to pass the exam and provide clients with sound advice.

Regarding the Series 7 Exam
Expect to be tested on your knowledge of these sophisticated and specialized investment possibilities and your ability to use that information in real-world situations. You should suggest acceptable investments for various customer categories or lay out complex ideas in clear language.

In finance, these sophisticated and specialized investment alternatives are your power plays, curveballs, and trick shots. They can provide investors with fresh opportunities to increase their wealth or mitigate risk, but they also come with challenges and considerations. This is only a brief overview; we'll learn more about them later in the chapter.

Delving Deeper into Security Investments
Now is the time to delve even further into the complexity of security investing. Understanding the intricate details of various investment vehicles can give you a significant advantage while you're preparing for the Series 7 exam or working to advance your financial advisory career. So, let's dissect it.

Derivatives
Financial contracts, known as derivatives, are those whose value is based on underlying assets like stocks, bonds, or commodities. They may be complicated but offer advanced risk management and hedging techniques. Options and futures are the two main categories to concentrate on. Explore more complex options trading techniques like spreads and straddles. Understand the commitment to buy or sell the underlying asset and operating margin requirements for futures.

Various Investments
Let's discuss many types of investments, including private equity, commodities, and venture capital. They are different from your standard possibilities and typically have larger risks and possible rewards. For instance, private equity may include buying out existing businesses, whereas venture capital may involve investing in start-ups. Whether gold or oil, commodities are another way to diversify a portfolio.

Complex Debt Securities

Consider alternatives to conventional bonds, such as zero-coupon, convertible, and collateralized mortgage obligations (CMOs). Each has its own set of restrictions, advantages, and disadvantages. For instance, zero-coupon bonds are bought at a significant discount and mature at face value but do not pay periodic interest.

Techniques for Portfolio Management

Utilize strategies like Modern Portfolio Theory and the Capital Asset Pricing Model (CAPM) to master the art of balancing risk and reward. Specialized and riskier investments are particularly crucial. For the Series 7 exam, you should also be familiar with rebalancing tactics and tax harvesting.

Tax Repercussions

Complex tax issues typically accompany advanced investments. Recognize the regulations governing tax-deferred accounts, short-term versus long-term capital gains, and the effects of various investment options on taxes. The better the net return for your client, the more you can optimize for taxes.

Risk Measures

Understand sophisticated risk indicators like beta (market risk), alpha (performance compared to the market), and the Sharpe ratio (risk-adjusted returns) to complete the picture. Calculate Using these criteria, you see the risks and possible benefits of sophisticated and specialized investment options.

You can anticipate seeing questions on the Series 7 exam that will test your comprehension of these complex instruments. You'll need to analyze data, formulate suggestions, and occasionally perform difficult computations instantly.

Borrowing Money and Securities: Margin Accounts

Prepare to plunge into margin accounts—a subject you must know for the Series 7 exam and your future as a financial advisor. The two-edged swords of the investment industry are margin accounts. Let's investigate how they operate since they can increase your gains and losses.

Describe the Margin Account

Investors can use a margin account to borrow funds from their brokerage to purchase securities. A margin account allows for leverage, in contrast to a cash account, where you can only invest funds that have been deposited. And leverage is the fuel for a financial rocket. It can either crash-land your capital or send your riches to the moon.

The Margin Request

The phrase "margin call" can be quite unsettling and understandably so. Suppose the value of your securities falls below a predetermined threshold known as the maintenance margin. In that case, your brokerage will issue a margin call. You may need to sell some assets or make a larger cash deposit to meet the required criteria. Unfortunately, your stockbroker may sell your securities without informing you, which can be painful.

Rates of Interest
Interest is an important consideration. Borrowing money has a cost, frequently a variable interest rate. The cost of borrowing and, ultimately, the profitability of your assets can be greatly affected by knowing the rate and how frequently it is capitalized.

Short Sales
Margin accounts also permit short selling, when shares are borrowed and sold to repurchase them at a loss. Remember that while you could earn if the stock price declines, losses could be limitless if it continues to rise.

Regulations and Threats
Great power entails enormous responsibility. Regulatory organizations have established Margin requirements to reduce some of the hazards. For instance, Regulation T of the Federal Reserve requires an initial margin requirement currently fixed at 50%. The exchanges and FINRA may each have their own set of regulations to follow.

Regarding the Series 7 Exam
Expect inquiries about margin computations, minimum standards, and the effects of a margin call. Understanding the mechanics and being able to run the figures can give you a big advantage not only on the test but also in practical applications.

There you have it—a thorough introduction to margin accounts. They are difficult, dangerous, and not appropriate for everyone.

Packaged Securities: Open- and Closed-End Funds
Let's solve the packaged securities puzzle, paying special attention to open-end and closed-end funds. Information on these sorts of investments is essential for the Series 7 exam and your future job as a financial advisor because they can fundamentally alter portfolio diversification and management.

Open-ended mutual funds
When people refer to "mutual funds," they typically mean open-end funds. These are noteworthy because, well, they're open. This indicates that shares may be issued and redeemed at any time at the Net Asset Value (NAV) determined after each trading day. Investors benefit from easy market access and professional asset management.

However, there is a catch: management costs, occasionally included in expenditure ratios, can reduce your returns. Additionally, if the fund is actively managed, you should be wary of turnover ratios because they can increase taxable events. For Series 7, be aware of how NAV is determined, the many fees involved, and the liquidity benefits.

Open-Ended Funds
These are open-end funds' less well-known and more enigmatic cousins. Through an initial public offering (IPO), closed-end funds issue a predetermined number of shares subsequently traded on exchanges like common stocks. This implies that their prices can fluctuate above or below the NAV depending on supply and demand. Cool feature one? They can use leverage to increase returns, but it also raises the risk.

Investors may find interest in particular methods or industries and the opportunity to purchase shares at a "discount" if they are selling below NAV. They also have management fees, and liquidity can be a problem.

Regulations and Threats
Both open-end and closed-end funds are subject to different regulatory requirements and dangers. Understanding the underlying assets and investment methods can assist in avoiding potential hazards, and due diligence is crucial.

Regarding the Series 7 Exam
It's important to be ready for questions that test your understanding of the differences between various funds, their operation, and the associated fees and risks. You may also encounter calculations involving NAV and discounts or premiums for closed-end funds.

Direct Participation Programs: Partnerships
Let me give you a brief introduction to Direct Participation Programs (DPPs) focusing on partnerships. Spoiler alert: These bad boys are a novel investing strategy that will appear on the Series 7 exam. Let's begin, then!

What are programs for direct participation?
To start, DPPs are not your typical assets like stocks or bonds. The underlying business initiative's cash flow and tax advantages are immediately accessible to investors through these investment arrangements. Although they can also involve other industries, these are frequently real estate or energy-related businesses.

GPs (General Partnerships) and LPs (Limited Partnerships)

Let's analyze the partnering scene now. General Partners (GPs) and Limited Partners (LPs) exist in a DPP. The venture's decision-makers, implementers, and brains are GPs. They are responsible for all obligations and debts. Conversely, LPs are more akin to silent partners; they contribute funds but have no input into everyday operations. However, they receive a portion of the earnings (and losses, just so you know).

Tax Advantages

The action now heats up. Partnerships immediately transfer profits, losses, and tax advantages to their partners. Depreciation and interest deductions could be a good bargain for your tax status if you invest in real estate. However, risk is once more a trade-off—especially for GPs who take on the most liability.

Cash Flow and Charges

Warning: DPPs aren't really that liquid. To pay out after you're in can be difficult. Additionally, various fees, including management and acquisition costs, are sometimes associated with these partnerships. Read the small print, then!

Regarding the Series 7 Exam

Here is your exam cheat sheet, all right. Be prepared to describe the functions of GPs and LPs and how they distribute earnings and losses. Know the tax repercussions because they are very significant. Not to mention the exorbitant cost structures and lack of liquidity, which may be deal-breakers for certain investors.

Options: The Right to Buy or Sell at a Fixed Price

Let's delve into the world of options, those fancy financial tools that grant you the right to purchase or dispose of an asset at a predetermined price. They can be complicated, yes, but don't worry! For Series 7 and your own knowledge, we will make it simpler.

What Kinds Of Choices Are There?

Options are agreements that give you the choice to buy or sell an asset, like a stock, at a set price, known as the strike price, within a certain period. There are two main types of options: call options, which grant the right to buy, and put options, which grant the right to sell.

Optional Calls

You gamble that the asset's price will increase when you buy a call. The stock could surge above your strike price, winning! Buying low and selling high gives you the opportunity. The premium you paid to buy that call is effectively gone if the stock doesn't reach the strike price before the option expires. In contrast, you want the asset to remain below the strike price if you sell calls to keep the premium.

Optional Puts

Puts make it seem as though you're betting against the spread. Your goal is for the stock to trade below the strike price. If so, you get the chance to sell high and purchase low. Lovely, no? When selling puts, however, you want the asset to remain above the strike price so you may keep the premium.

Greek Script (No, Really)

Options traders frequently use Greek words like "Delta," "Gamma," and "Theta." These assist in determining the risk and potential reward. Knowing these can give you an advantage when trading and taking the Series 7 exam.

Benefits and Risks

Options give leverage, allowing you to control many shares for a very small investment, making them excellent for hedging against losses in other assets. The danger, though? If the option expires worthless, you could lose everything you invested. Be cautious hence.

Hot Takes for Series 7 Exam

Expect inquiries regarding the fundamentals, such as what calls and puts are, their dangers, and how they might be utilized for hedging or speculation. There may also be computations, including premiums, gains, and losses.

PART IV: TRANSACTION HANDLING AND COMPLIANCE

Chapter 10

Transaction Management

We will start Chapter 10, which is all about transaction management. This is an important question, especially if you are preparing for the Series 7 exam. It encompasses everything, from the execution of trades to the validation of transactions and account statements. To be at the top of your game in the financial industry, it is essential to have a solid understanding of how transactions are handled. Let's dive right in, shall we?

The receipt of the order
A blueprint is a written record of what needs to be done, and this document serves as the blueprint for any trade. It contains the customer's name, the type of order to be placed, the security that will be traded, and other data. If you botch this, then everything will fall apart like dominoes. On the Series 7 exam, you can be asked questions evaluating your knowledge of order tickets.

Different kinds of orders
You can place Market Orders, Limit Orders, and Stop Orders; each of these three types of orders comes with its own distinct rules and parameters. Not only is it necessary to pass Series 7, but also for successful transaction management in real life. Knowing when to employ which is gold.

Execution of Trades
This is the point when the proverbial rubber meets the road. After an order ticket is completed, the order will either be sent to the floor of the exchange or to a trading platform to be executed. In this context, time equals money, and delays can be expensive. Understand the various paths an order might take, such as passing through a broker-dealer or proceeding straight through an electronic system.

Confirmations and Agreements on Terms and Conditions
Therefore, your trade was completed. Very nice! The next step is confirmation, which entails sending a document that describes the entire transaction in great detail. It needs to be checked, double-checked and then checked once more. Any mistakes can result in significant difficulties further down the line.

After that comes the settlement, which is the real trading of money and securities for one another. There is a possibility that the questions on Series 7 will need you to demonstrate an understanding of the settlement procedure, including settlement dates.

Financial Reports and Statements
These bad boys provide an overview of the account, detailing everything that has been performed, such as transactions, cash balances, and security positions. It is similar to the "Previously on…" recap shown at the beginning of each television show episode, and it brings you up to speed on what has been going on.

The Value of Obeying the Rules and Regulations
The purpose of regulations is not only for show; they are intended to ensure that everyone plays by the rules. Because failing to comply might result in severe penalties, you should anticipate seeing "compliance" on the Series 7 form. Get a good understanding of the important documents that should be kept and for how long.

This concludes your introduction to the administration of transactions. Suppose you are familiar with all of this information to the point that it is second nature. In that case, you will be prepared for the Series 7 exam and understand how the trading world's back-office functions.

Providing Current Quotes
This may appear elementary, but believe me when I say it is really important, regardless of whether your objective is to ace the Series 7 or simply to be an expert at your work. Let's get down to business here!

What are the Latest Quotations?
Think of the stock market's mood ring like a quote that shows a stock's current buying and selling sentiment. A quote includes the last traded price, the highest price someone is willing to pay (known as the bid), and the lowest price someone is willing to sell (known as the ask). These topics may arise in Series 7, so be prepared for questions about them.

Comparison of Real-Time and Delayed Quotes
Quotes that are delayed by a few minutes provide less current information than those provided in real time. You may be required to discriminate between them on the Series 7, so keep this in mind.

The Process Behind Obtaining Quotes
Brokers can obtain these quotes from various sources, including web services, direct data feeds, and trading platforms. But the most important thing is to understand how to decipher them. It is a huge

no-no to misread a quote because this can lead to the execution of a trade at a price that was not intended.

Quotes that are both Indicative and Firm
This is where things get a little more technical. Indicative quotes will provide a rough cost estimate but are not final. On the other hand, firm quotations are rates that can be transacted, and that dealers and brokers must honor. You need to be aware of the distinction to succeed in real-world trading and, yes, in the Series 7 exam.

Influence of the Current Market Conditions
Various external factors can influence A quote's value, including market news, economic data, and even public opinion. If you are a broker, more is needed to provide the price; you must also provide some insight into why the quote is what it is.

Why It Is So Important for the Series 7
It is crucial to have the ability to deliver accurate quotations. Use your ability to interpret various quotes, comprehend the elements that influence them, and demonstrate an understanding of how they fit into a bigger market strategy.

Both compliance and documentation are required
When it comes to transaction management, having a paper trail is quite essential. Brokers must retain records of the quotes they issue to their customers, particularly if they result in a trade being executed. This is more than merely considered good practice; it is typically required by various regulations.

Providing current quotations requires more than merely reciting a series of numbers; it requires a comprehension of what those numbers signify, how they function within the context of the larger market picture, and how to use them.

Processing and Confirming Customers' Transactions
Let's discuss "Processing and Confirming Customers' Transactions" in the context of Transaction Management. This topic is extremely significant, whether you are getting ready for Series 7 or simply want to better grasp what goes on behind the scenes. Therefore, let's go to work!

The Path from the Order to the Punishment
Executing the trade is the first step in the real action, which comes after creating the order ticket and completing all the crucial preparation steps. This step may send the order to the trading floor, an electronic communication network, or a market maker. Because time is of the essence, it is essential to execute the plan quickly and correctly. You may be sure that the Series 7 exam will test your knowledge of the more nuanced aspects of order execution!

Confirmation Following the Act of Execution

A confirmation slip is prepared and mailed to the customer when the order is processed and fulfilled. This document serves effectively as a receipt, as it details all the aspects of the transaction, such as what was bought or sold, the price, the date, and any fees that may have been applicable. This needs to be checked more than once to ensure accuracy. Errors are not just a potential source of financial loss but also of regulatory hassles waiting to come.

Maintenance of Records and Observance of Regulations

And while we're on the subject of regulations, we must keep precise records. This entails maintaining a record of the confirmations and initial order tickets, any modifications made to the orders, and all conversations with customers. Be prepared for the Series 7 exam to throw a curveball at you by asking about compliance processes for record-keeping; it may do so.

It's Time to Settle Down!

The actual handing over of assets is what is meant by "settlement." This normally occurs two business days after the deal has taken place in the United States for most securities (T+2). It is important to take advantage of this step; failures or delays in settling the account can result in financial penalties. Watch out for Series 7 inquiries that ask you about the timing and methods of settlements; they will likely come up.

Reconciliation, as well as Statements for Customers

At the end of the business day (or more likely once a month), the client receives an exhaustive account statement that provides a summary of all operations, including trades, dividends, interest, and any other transactions that have taken place. Another essential piece of paperwork that needs to be perfect is this one.

Why This Is Important Regarding the Series 7

You should be prepared to answer questions covering the entirety of a transaction's lifetime, from the placement of an order to its confirmation to its finalization. It is essential, not only for the exam but also for your future career in finance, that you have a solid grasp of the mechanics and the regulatory criteria at each stage.

Addressing Margin Issues

Let's begin by discussing "Addressing Margin Issues" in the context of Transaction Management. If you are getting ready to take the Series 7 exam, or even if you are curious, margin issues are a topic you will want to thoroughly understand. Should we break it down into its component parts?

A Primer on Margin

To begin, let's get down to the fundamentals: Can you explain what a margin account is? You can open an account with a broker that allows you to borrow money from them to buy securities and use the securities you already possess as collateral for the loan. You will be expected to comprehensively understand this topic, including the minimum deposit requirements and maintenance margins. Using the margin can increase your purchasing power, but doing so exposes you to greater financial risks and expenses. Therefore, watch your step!

The Pressure Is On

There will probably be questions on this topic in Series 7. A margin call happens when the assets in a margin account decrease below the maintenance margin. The broker will then ask for more cash deposited into the account. Failing to satisfy a margin call may result in the broker liquidating the client's shares to cover the shortfall in value. It is imperative to steer clear of a "fire drill" situation.

The T Regulation

You'll need to have this knowledge for Series 7 as well, that's for sure. The first margin requirement is established by Regulation T. This requirement specifies the proportion of the purchase price that must be deposited. The most recent information indicates that it was 50%, but since rules and regulations are subject to change, you should watch for the most recent data.

Rules Regarding House Calls and Day Trading

Many brokerage houses have their own internal set of margin requirements, called "house calls," in addition to the "Reg T" calls. There is a good chance that Series 7 will test your knowledge of the distinctions. Please be aware of the regulations around day trading. The margin requirements will change if you execute four or more trading deals in a single day within five days. This will classify you as a "Pattern Day Trader," and a higher minimum equity requirement will be imposed on your account. It's important to keep this in mind when managing your trades.

Transactions

Margin requirements make things more complicated for transaction management professionals. When you're in the middle of making trades and all of a sudden get a margin call, it can put a stop to or change your trading methods. Because of this, keeping a tight eye on margin accounts is of the utmost importance. Real-time margin monitoring is a tool offered by many trading platforms, and it can be used to help avoid margin calls. However, there is no substitute for the consistent hard work of humans.

The Angle of the Series 7

For the Series 7 exam, a thorough understanding of margin concerns is necessary. You should expect to encounter inquiries on various topics, including initial margins, maintenance margins, margin calls, and regulatory obligations. Consequently, add this to your list of things to study!

Settlement Procedures

When it comes to transaction management, having a solid understanding of settlement procedures is crucial, and this is especially true if you are preparing for the Series 7 exam. Not only will it impact the general flow of your transactions, but it will also be one of the topics that the Series 7 exam will test your knowledge of. Now, let's get down to the nitty-gritty of this subject.

The Fundamentals: T+1, T+2, and Beyond are All Beyond

First, let's discuss the letter "T" in the phrase "settlement procedures." The letter "T" denotes the transaction date, which is the day on which you or a client of yours clicks the "Buy" or "Sell" button. In the past, the settlement date for many assets, such as stocks and bonds, was T+3 (meaning three days after the transaction), but as of the most recent information that I have, it is almost always T+2. Keep an eye out for the most recent settlement periods in the Series 7 preparation materials you've acquired, as they are subject to change. You must familiarize yourself with the specifics of the various settlement periods that may apply to mutual funds and other types of assets.

Contrasted with margin accounts are cash accounts

The kind of account you trade from can make a difference in the settlement process. Settlement does not take place in a vacuum. When using a cash account, you must have the total amount of the purchase available when the transaction is settled. You must meet the minimum initial margin requirement to open a margin account. Series 7 will demand you to be aware of the changes and what will occur if you cannot settle. The bad news is that none of it is any good. You get the picture: penalties for being late and frozen accounts.

The Availability of Funds

You are responsible for being informed about the time frame within which the proceeds from a sale can be withdrawn or used to invest in additional securities. You may be asked about this on the Series 7 exam, so you should pay attention to it. In most cases, you will fully possess the funds one day following the settlement date; certain brokerage firms may have additional regulations.

Risks Involved in Settlement

Okay, here comes the "be careful" section of the sentence. If, for whatever reason, either the buyer or the seller cannot finish their portion of the transaction by the settlement date, then we will have to deal with the consequences of a failed settlement. This can lead to penalties and mess up the trading plan you devised for yourself.

Confirmations of Transactions and the Maintenance of Records
After the deal has been finalized, you will often receive an email confirmation that outlines the transaction. It is not merely a receipt; rather, it is an essential document for the purposes of record-keeping and regulatory compliance. You'll want to hang onto those for a while, given the tight laws in the banking industry, particularly when you're functioning in the capacity of a registered representative.

You must be familiar with your settlement procedures. On the Series 7 exam, you will be asked questions on this topic, and in your day-to-day work, you will need to know it like the back of your hand. You can keep track of how and when transactions settle. In that case, you can keep everything running smoothly and avoid unpleasant surprises.

Chapter 11

Compliance, Rules, and Regulations

Compliance, rules, and regulations are perhaps not the most exciting aspect of the financial sector but trust me when I say that ignoring this topic is not an option, especially if you're studying for the Series 7 exam. The information that you must have can be found below, divided up into manageable chunks:

The Letters of the Alphabet in Regulation
FINRA, SEC, MSRB, and SIPC are just a few of the many acronyms that are used in the financial business. You must be familiar with these individuals and the work that they conduct. For instance, the federal agency responsible for supervising compliance with securities regulations is the Securities and Exchange Commission (SEC). The Financial Industry Regulatory Authority, or FINRA, is an example of a self-regulatory agency that contributes to enforcing these rules. Acronyms will be on the test and your everyday job life, so you should get familiar with them.

Learn Your Own Acts
You should be familiar with the primary laws that govern the securities sector, starting with the Securities Act of 1933 and moving up to the Investment Advisers Act of 1940. Every act has its rules, but to keep things interesting, they frequently overlap with one another. The Series 7 exam will evaluate your understanding of these aspects of securities regulation, which are the exam's foundational pillars.

Morality and Social Conduct
Let's cut to the chase: ethics play a significant role in our lives. The Financial Industry Regulatory Authority (FINRA) has a set of ethical norms known as the Rules of Fair Practice, and financial professionals must comply with them. We are conversing about offering appropriate recommendations, being open and honest about the expenses, and not attempting to manipulate the client's order. It is essential to have a comprehensive understanding of the rules, as the slightest deviation from them can result in severe punishments or even the suspension or revocation of your license.

Conversations with members of the public
It is important to remember not to undervalue the power of words, particularly when it comes to advertising and connecting with customers. When discussing different investment opportunities, some rules govern what you can say, write, or even imply. Misleading advertisements or empty pledges? It's a big no-no. You will be questioned extensively about what you are permitted to say and how you are supposed to say it when you participate in Series 7.

Suspicious Activity Reports and Measures to Prevent Money Laundering
Guys, take this discussion seriously. You are responsible for identifying and reporting any suspicious actions, as the institutions you deal with have implemented severe Anti-Money Laundering (AML) regulations. Consider the unexpected appearance of enormous sums of money in accounts that cannot be explained, as well as questionable wire transfers. Because you are on the front lines, you are responsible for taking action if you observe questionable behavior.

Keeping Records and Producing Documentation
You are mistaken if you work in the financial industry and believe you can avoid completing paperwork. Keeping accurate records is essential to regulatory compliance, and this is true for more than simply the auditing process. You have to organize and archive everything to make it easy to find, whether it be information about customers, histories of transactions, or internal communications. It is a requirement and a topic discussed frequently on the Series 7, even though it may sound tedious.

Public disclosures
Let's discuss disclosures as our very last item on the agenda. Each of the several forms that clients must sign, and the account forms and risk documents contain a compliance component. Provide the necessary disclosures to avoid the creation of legal issues and a loss of trust. You are correct in assuming that the Series 7 examination would also test your knowledge of this topic.

Taking Care of Your Customers and Playing by the Rules
Compliance, rules, and regulations are perhaps not the most exciting aspect of the financial sector but trust me when I say that ignoring this topic is not an option, especially if you're studying for the Series 7 exam. The information that you must have can be found below, divided up into manageable chunks:

The Letters of the Alphabet in Regulation
FINRA, SEC, MSRB, and SIPC are just a few of the many acronyms that are used in the financial business. You must be familiar with these individuals and the work that they conduct. For instance, the federal agency responsible for supervising compliance with securities regulations is the Securities and Exchange Commission (SEC). The Financial Industry Regulatory Authority, or FINRA, is an

example of a self-regulatory agency that contributes to enforcing these rules. Acronyms will be on the test and your everyday job life, so you should get familiar with them.

Learn Your Own Acts
You should be familiar with the primary laws that govern the securities sector, starting with the Securities Act of 1933 and moving up to the Investment Advisers Act of 1940. Every act has its rules, but to keep things interesting, they frequently overlap with one another. The Series 7 exam will evaluate your understanding of these aspects of securities regulation, which are the exam's foundational pillars.

Morality and Social Conduct
Let's cut to the chase: ethics play a significant role in our lives. The Financial Industry Regulatory Authority (FINRA) has a set of ethical norms known as the Rules of Fair Practice, and financial professionals must comply with them. We are conversing about offering appropriate recommendations, being open and honest about the expenses, and not attempting to manipulate the client's order. It is essential to have a comprehensive understanding of the rules, as the slightest deviation from them can result in severe punishments or even the suspension or revocation of your license.

Conversations with members of the public
It is important to remember not to undervalue the power of words, particularly when it comes to advertising and connecting with customers. When discussing different investment opportunities, some rules govern what you can say, write, or even imply. Misleading advertisements or empty pledges? It's a big no-no. You will be questioned extensively about what you are permitted to say and how you are supposed to say it when you participate in Series 7.

Suspicious Activity Reports and Measures to Prevent Money Laundering
Guys, take this discussion seriously. You are responsible for identifying and reporting any suspicious actions, as the institutions you deal with have implemented severe Anti-Money Laundering (AML) regulations. Consider the unexpected appearance of enormous sums of money in accounts that cannot be explained, as well as questionable wire transfers. Because you are on the front lines, you are responsible for taking action if you observe questionable behavior.

Keeping Records and Producing Documentation
You are mistaken if you work in the financial industry and believe you can avoid completing paperwork. Keeping accurate records is essential to regulatory compliance, and this is true for more than simply the auditing process. You have to organize and archive everything to make it easy to find,

whether it be information about customers, histories of transactions, or internal communications. It is a requirement and a topic discussed frequently on the Series 7, even though it may sound tedious.

Public disclosures

Let's discuss disclosures as our very last item on the agenda. Each of the several forms that clients must sign, and the account forms and risk documents contain a compliance component. Provide the necessary disclosures to avoid the creation of legal issues and a loss of trust. You are correct in assuming that the Series 7 examination would also test your knowledge of this topic.

Rules and Regulations: No Fooling Around

There is no space for playing fast and loose in finance regarding adhering to compliance standards, rules, and laws. No quick cuts, no ambiguous areas, simply straightforward research and preparation. Let's go down the reasons why this is such an important point.

Straight to the Point from the Beginning of the Process

After successfully completing the Series 7 exam, you will be a part of a highly regulated world, right down to the smallest details. The purpose of these regulations is not to make your life more difficult; rather, they are in place to safeguard customers and ensure that the integrity of the financial markets is preserved. Therefore, it is not open for discussion whether one must comply with the requirements of the SEC or the rules of FINRA. Should you fail to comply, disciplinary sanctions, significant penalties, and even criminal charges may be taken against you. Believe me when I say that you do not want any of that gloomy business showing up on your doorstep.

The Watchdogs Are the FINRA and the SEC

Regarding regulatory monitoring, the Financial Industry Regulatory Authority (FINRA) and the Securities and Exchange Commission (SEC) are the two most important players in the game. They have many rules and guidelines meant to assure fairness and openness in the market, and they have those rules and principles in place. It is important to remember that these organizations not only act as gatekeepers but also as referees. These organizations will sound the alarm if you step in the wrong direction.

Due Diligence on the Part of the Customer Get to Know Your Customers

The "Know Your Client" (KYC) guideline is one of the most important compliance pillars. It's not simply an old-fashioned sales pitch; it is a regulatory compliance requirement. Before offering any advice, financial advisors must first understand their customers' financial objectives, level of comfort with risk, and previous investment experience. This is about more than simply providing outstanding customer service; it's about ensuring that you are behaving in the best interests of your clients and not putting them in situations where they are exposed to hazards that are not required.

Buying and Selling and Reporting

Trading in securities is a highly regulated business that is subject to a wide variety of guidelines. Trading hours, price reporting, and the execution of trades are all subject to certain rules. Trading with inside information while also trying to manipulate the market? That is a major no-no in my book. These behaviors constitute criminal violations, and violators could face significant repercussions.

Keeping Records and Producing Documentation

It is impossible to stress the significance of keeping accurate records appropriately. The maintenance of accurate records is an essential component of regulatory compliance. These records may pertain to interactions with customers, the specifics of transactions, or disclosure paperwork. In the event of an audit or investigation, these records may likely need to be provided; if you do not retain them, it could be disastrous for both you and your company.

Potentially Conflicting Interests

The way things should work is that your interests and customers' interests should never compete with one another. Do not engage in self-dealing, do not accept bribes, and under no circumstances should you ever promote investments in which you or your company have a financial interest unless such recommendations have been properly disclosed and agreed upon. Everything that resembles a conflict of interest is likely to attract the notice of regulatory authorities, and you are required to handle situations involving conflicts of interest with the highest honesty.

Training and Supervision on an Ongoing Basis

Maintaining compliance with all laws and regulations is not a task that can be completed in a single sitting. It's a lifelong dedication to the cause. Because of this, it is essential to participate in continual compliance training and continuing education classes. In addition, many businesses have internal compliance teams tasked with monitoring activities and ensuring that everyone is acting according to the regulations.

Protection against cyberattacks

Compliance with cybersecurity standards is becoming increasingly important in a growing increasingly digital society. The prevention of illegal access to customer data is an issue that pertains not just to information technology but also to regulatory compliance. Failure to comply might result in losses not just in terms of money but also in terms of reputation.

Because of this, whenever we discuss policies and procedures, we do so in a solemn manner since the stakes are extremely high.

PART V:
TAX CONSIDERATIONS AND RETIREMENT PLANNING

Chapter 12

Taxes and Retirement

Taxes and retirement are two crucial subjects that every prospective financial counselor or someone with a Series 7 license must have a solid grasp on. A comprehensive understanding of these topics, even though they could appear dull or even bewildering to you, is necessary to develop your professional skills and provide useful financial guidance to your customers. I'll explain why.

The Unavoidable Consequence: Paying Taxes
Although no one enjoys doing so, paying taxes is one of the few things that can be counted on in life, as Benjamin Franklin famously said. Although you won't be responsible for drafting tax returns as a financial advisor, you will still need to know the tax repercussions of various investment opportunities. It is essential to have a solid understanding of the potential impacts that taxes, such as those on capital gains and income, as well as tax-free investment opportunities, might have on a client's portfolio. The knowledge you possess helps your customers save money and can elevate you from being a competent advisor to a great one.

Gains on investments and dividends both count
It is common for customers to make investments without considering the potential tax consequences of their choices. For example, the time an investment is kept might result in varying tax rates being applied to various types of income, such as dividends and capital gains on stocks. Because the rate of taxation on gains realized in a short period is typically higher than that on gains realized over longer periods, it may be only sometimes the greatest option to sell an asset simply because its value has increased.

Investments that are Either Tax-Deferred or Tax-Exempt
Products such as 401(k)s and Roth IRAs offer a variety of tax advantages, which, when included in a diversified investment portfolio, can prove helpful. When taxes are deferred or eliminated entirely, the difference in compounded returns over several years might be quite large.

Bonds and municipal issues

The purchase of tax-exempt municipal bonds, sometimes known as "munis," is a fundamental component of any tax-advantaged investment strategy. Even though they may provide a smaller yield than taxable bonds, their tax savings can result in a higher effective yield, especially for those with higher tax rates.

The Long Game: Getting Ready for Retirement

Let's face it: everyone has the same goal: to retire with enough money to live comfortably in their golden years. As a financial advisor, one of your key obligations is to assist your customers in achieving this objective. It's not enough to merely suggest a few different retirement accounts; you need to develop an all-encompassing strategy that considers your comfort level with risk, the length of time you have to invest, and the market volatility.

IRAs and 401(k)s

Here are the key components to consider when planning for a successful retirement. Individual Retirement Accounts (IRAs) are personal bank accounts, while 401(k) plans are typically set up by employers. Both offer tax benefits but in different ways. For instance, some 401(k) plans match employer contributions dollar for dollar, essentially free money that should be noticed. On the other hand, IRAs offer more investment options and flexibility but usually do not include employer contributions or matching funds.

Variations on Roth

Roth 401(k)s and Roth IRAs differ from traditional retirement accounts because contributions are made after taxes are paid. However, no taxes will be deducted from distributions once you retire. To effectively manage your clients' retirement portfolios, it's important to understand the advantages and disadvantages of both options.

Plans for Retirement and Annuities

These occur less frequently yet are nonetheless essential to comprehend. They provide a consistent income throughout retirement, but they come with their own laws and regulations that must be followed.

The Social Security System

Even though you are not a Social Security advisor, you must understand how the benefits provided by Social Security fit into your client's retirement plan. For instance, there are tactical considerations to consider while deciding whether to begin receiving benefits at the earliest age possible of 62 or to wait until as late as 70 to obtain the highest potential monthly payout.

Inheritance Tax Planning and Estate Planning
Lastly, regarding retirement planning, it's important to consider not just the clients themselves but also their heirs and how they will be taken care of in the coming years. Estate planning is a crucial aspect of financial planning that deserves more attention. It's vital to have a good grasp of the tax implications of bequests and to understand how trusts and other estate planning techniques can be used to minimize the amount of tax owed on an inheritance.

The Comprehensive View
Suppose you are extremely knowledgeable about tax regulations and retirement plans. In that case, you will be an extremely valuable addition to any customer you work with. You'll have the ability to provide individualized guidance that can save them money in the short and long term, as well as provide them the piece of mind that comes from knowing they're on the proper path toward a comfortable retirement. However, even though taxes and retirement planning aren't the most exciting topics, understanding them is vitally necessary for the work you'll do as a financial advisor. If you can get a handle on these, you'll be well on your way to having a successful career.

Taxes and Retirement Plans: Making Sure the IRS Gets Its Share
You exercise extreme caution to ensure the IRS receives its due without causing your customers to pay more than they should.

Taxes and Retirement: Striking the Right Balance
The process of paying taxes and preparing for retirement can be compared to a delicate dance in which the taxpayer and the Internal Revenue Service (IRS) are required to make strategic moves. Tax regulations determine how retirement funds are handled, and depending on the circumstances, investors may either profit or suffer due to these regulations. Financial advisors need to comprehensively understand the complex tax regulations regarding retirement plans. This helps to create effective and tax-efficient investment strategies.

The Current Tax Climate Traditional Individual Retirement Accounts and 401(k) Plans
Many people contribute to tax-deferred retirement accounts, like Traditional IRAs and 401(k)s, before taxes are deducted from their account balance. This is a wise decision as the total amount will benefit from compound growth over time. However, when the funds are withdrawn during retirement, they will be taxed like any other income source. Therefore, it's important to consider the potential tax bracket that your customer may be in after retirement.

Accounts such as Roth IRAs and Roth 401(k)s
These accounts are funded with money that has already been subject to taxation. Therefore, any subsequent withdrawals are exempt from taxation. Explain the importance of tax-free growth and

withdrawals. You can convince younger clients to pay tax now in exchange for a benefit in the future. However, this may take a lot of work to accomplish.

Plans for Retiring That Are Not Considered Qualified

Plans like annuities and various kinds of deferred compensation plans are examples of these types of plans. The tax ramifications are more complex to understand, even though they provide diverse options for income and development. Earnings may be allowed to grow tax-deferred in certain circumstances; however, the rules can take time to follow.

The Predicament of Withholding

When collecting distributions from their retirement funds, many individuals are under the impression that they cannot—and indeed, should not—specify their preferred tax withholding method. When tax season comes around, failing to withhold an adequate amount of money might result in underpayment penalties. Your advice can be quite helpful in this situation.

The Pitfall of the Early Withdrawal

Precise guidelines must be followed to avoid incurring penalties when taking money out of retirement plans such as 401(k)s and IRAs. Usually, the account holder must wait until they are 59 and a half years old before getting payouts. A 10% early withdrawal penalty is added to the usual income tax if this does not occur. The penalty can be an unpleasant surprise, but it can be prevented with adequate planning and by educating the customer.

RMDs are abbreviations for "Required Minimum Distributions."

When customers reach the age of 72, they must start withdrawing a certain amount from tax-deferred retirement funds called required minimum distributions (RMDs). This amount is taxable income and is determined by your life expectancy and the balance in your account. Ignoring required minimum distributions (RMDs) can result in a tax penalty of up to fifty percent of the amount that should have been withdrawn, which no customer wants to experience.

Exceptional Fiscal Laws and Tax Credits

A worker with a low or moderate income who makes contributions to a retirement account may be eligible for the Saver's benefit, which is a tax benefit. It is crucial to be aware of this credit. In addition, customers still working after 72 can suspend required minimum distributions (RMDs) from their current employer's 401(k) until they retire. This is a distinct treatment compared to IRAs, which require RMDs to be taken out annually.

Diversification of taxes

Tax diversification is a topic that is frequently ignored, and it refers to the practice of maintaining assets in many types of accounts to manage tax risk. Suppose you have a combination of taxable, tax-deferred, and tax-exempt accounts, for instance. In that case, you have more leeway when deciding how to distribute your money after retirement, which helps you more effectively manage the burden of paying taxes.

Important Tax and Estate Planning Considerations

Your customer isn't the only one who will be impacted by these tax ramifications; their beneficiaries will also. You can learn about inheritance regulations and the tax treatment of various accounts. In that case, you will be able to assist your customers in formulating an estate plan that will reduce the amount of money owed in taxes by their heirs.

Regulatory Shifts and Continuous Education and Training

Tax regulations rarely always remain the same. It is crucial to stay current with these developments to provide the most appropriate advice, as they are subject to change in response to newly enacted legislation. The SECURE Act and other legislative reforms could bring about a swift shift in how retirement accounts are regarded for taxation purposes.

The Closing Statement

A vital service that adds layers of complexity to a landscape that is already difficult is assisting customers in navigating the intricate tax consequences of the many retirement plans that are available to them. You must play a part in ensuring that the Internal Revenue Service receives its due while also ensuring that your customers only pay a penny as much as they have to. You may save your customers money and alleviate much worry by remaining knowledgeable about the tax rules and how they interact with retirement planning. This will further establish your value as an invaluable financial advisor.

PART VI:
FINAL PREPARATIONS AND EXAM TIPS

Chapter 13

Portfolio and Securities Analysis

Analysis of a portfolio's holdings and its securities, both in art and science, should be considered essential abilities for any financial advisor. The success or failure of an investing plan can be directly correlated to the level of understanding of the underlying mechanisms, dangers, and potential for returns, as well as the interconnections between various asset classes and particular securities. To successfully pass the Series 7 exam, you will need to comprehensively understand how to analyze a portfolio as a whole and its various components. Let's get right into some fundamental ideas that will guide you as you move forward with this process.

Asset Distribution: The Fundamentals of a Strong Foundation
The process of allocating assets involves more than simply spreading investments over various asset classes; rather, it entails selecting the optimal combination of assets to accomplish particular goals while considering the level of risk tolerance a customer possesses. For instance, a younger customer who is willing to take on a greater amount of risk might have a greater allocation towards stocks, whereas an older client might have a portfolio that is more heavily weighted towards bonds and other fixed-income instruments.

The Power that Comes from Having a Wide Variety of Options
Diversification is not only a trendy concept; it is a principle founded on the empirical observation of mathematics. A portfolio can earn a higher return for a given level of risk if the assets are diverse and not highly linked. To put it another way, a diversified portfolio allows for the possibility of achieving a certain level of return while simultaneously lowering the associated level of risk. Graphically depicting the advantages of diversification, the efficient frontier and the capital market line are the tools that illustrate the concept of the efficient frontier and the capital market line.

Evaluating equities can be done through quantitative and qualitative approaches.
Financial advisors have various instruments for conducting research and analysis regarding stocks and shares. Quantitative measures like Price-to-Earnings ratios, Debt-to-Equity ratios, and others provide a

numerical overview of a company's overall financial health. In the meantime, qualitative aspects such as the quality of management, trends in the industry, and market position provide additional layers of knowledge that can't be captured by numbers alone.

Fixed Income Securities: Considerations Other Than Yields
At first glance, bonds could appear to be a less thrilling investment option than stocks. They are typically regarded as having lower volatility levels and provide fixed yields. However, the examination of bonds is a complicated process that incorporates a variety of considerations, such as credit ratings, the sensitivity of a bond to changes in interest rates (which are measured by the bond's term), and the issuer's current financial status. Because the many kinds of bonds, such as Treasury, municipal, and corporate bonds, have unique risks and tax implications. Prior knowledge is critical before investing in any bond.

Options, Futures, and Other Derivatives: Products for Experienced Investors
The study of a portfolio is made much more complicated by including instruments such as options and futures. It is absolutely necessary to have a thorough understanding of their reward structures, the risks that are associated with them, and how they connect with the other assets in the portfolio. They can be used for various things, including protecting oneself from risk or increasing one's position through leverage.

Properties Available and Other Investment Opportunities
A diversified portfolio can also comprise real estate and alternative investments like hedge funds or commodities. These assets have different risk and return profiles than traditional stocks and bonds, giving investors more diversification options. However, they also have unique challenges, such as illiquidity and higher expenses, which may make them less attractive.

An Examination of Risk Using Variance and Standard Deviation
Calculating the variance and standard deviation of asset returns is an integral part of the scientific approach to portfolio analysis. These indicators contribute to a better understanding of the degree of volatility shown by individual securities and the portfolio as a whole. A portfolio with a high standard deviation might yield higher returns, but it will also come with larger risks, making it inappropriate for investors who prefer to minimize their risk exposure.

The Beta Coefficient Is Used To Determine How Sensitive The Market Is
While variance and standard deviation are extremely important, their information must be more comprehensive. Beta is another important indicator that assesses the degree of a security's volatility compared to that of the market as a whole. A security's beta that is larger than 1 implies that its volatility is higher than that of the market, while a beta that is less than 1 indicates that its volatility is

lower than that of the market. When building a portfolio for a customer, it is helpful to know the beta of each individual asset so that the portfolio can be structured to accommodate the client's risk tolerance.

A Key Concept in Contemporary Portfolio Theory: the Efficient Frontier
The Modern Portfolio Theory (MPT), developed by Harry Markowitz, fundamentally altered how people conceive of and approach investments. The Modern Portfolio Theory explains that more is needed to look at a single investment's anticipated risk and return. According to the idea, it is important to consider how each investment performs compared to the performance of all the other investments in the portfolio. The ideal portfolio does not offer the highest possible expected return for a given level of risk; rather, it gives the lowest possible risk for a given expected return. This is because the optimal portfolio seeks to maximize returns while minimizing risk.

Analysis Based on Fundamentals as Opposed to Analysis Based on Technicals
Fundamental and technical analysis are the two basic schools of thinking predominating security analysis. A company's financial records and health, its management, and its advantages over its competitors are only a few of the factors taken into consideration throughout the fundamental analysis process. On the other hand, to estimate future price changes, technical analysis emphasizes statistics created by market activity. These statistics include things like past prices and volume. Both have benefits and limitations, so many advisors combine the two strategies.

Management That Is Active as Opposed to Passive Management
Both of these investment approaches have the potential to drastically affect the performance of a portfolio. Active management is when individual investments are selected to exceed a benchmark index that measures the performance of investments. On the other hand, passive management's objective is to replicate the composition of investment assets contained within a specified benchmark index. The dispute between active and passive voice is one that you are sure to come across, and it is one that you will need to have a comprehensive understanding of to provide useful advice to your clients.

Exchange-Traded Funds and Mutual Funds
Exchange-traded funds, also known as ETFs, and mutual funds both allow investors to buy a high level of diversity with a single transaction. Retail investors, who may need more financial means to purchase a diversified portfolio of individual stocks and bonds, may find this a perfect investment option. Your study of your portfolio can benefit from having an additional layer of depth if you have a solid understanding of the differences between these two.

Rebalancing one's portfolio is an important part of maintaining control
Reviewing and rebalancing a client's portfolio regularly helps to ensure that it continues to reflect the client's original investment objectives. The asset allocation of a portfolio can be shifted, and as a result, it can become either too risky or too conservative. This can be caused by several different life events or market situations. It is necessary to have the ability to recognize when rebalancing is necessary and how to do it.

Statement of Investment Objectives and Objectives for Investment: The Blueprint
When making decisions about a portfolio, having an Investment Policy Statement (IPS) can act as a useful strategic guide. The Investment Policy Statement (IPS) serves as a roadmap for the investment process while also providing an account of the client's financial goals, risk tolerance, and investment objectives.

Cognitive Fallacies and the Psychology of Investing
The psychological state of an investor can have a substantial impact on the returns their portfolio generates. Financial advisors can better steer their clients toward making more reasonable decisions by understanding typical behavioral biases such as loss aversion, overconfidence, and herd mentality.

There is no room for negotiation when it comes to ethics and compliance
A non-negotiable component of portfolio management is ensuring that ethical considerations are considered and remain by applicable rules. Protecting yourself and your customers requires maintaining a high level of openness, following your fiduciary responsibilities, and keeping current on the latest applicable laws and regulations.

Analysis of portfolios and securities is a multidimensional field that covers a wide variety of subjects, from the most fundamental aspects of asset allocation to the most complex financial theories and ethical considerations. Having a firm grasp on these will prepare you for the Series 7 test and provide you the tools you need to construct and manage portfolios tailored to the specific requirements and risk profiles of each of your clients.

Examining Companies and the Market
When it comes to trading on the stock market, the intricacies of a trade are often where you will find the most difficulty. Candidates for the Series 7 exam, including prospective investors and financial advisors, must demonstrate a mastery of the art of company and market analysis. This entails analyzing the financial statements of a company, becoming familiar with the market dynamics, and taking macroeconomic data into consideration. A comprehensive look at the most important aspects is presented here.

Acquiring Knowledge of Financial Statements

Suppose you're considering recommending a stock to an investor. In that case, it's essential to carefully examine the financial statements of the company in question. This involves analyzing the key components of the income statement, balance sheet, and cash flow statement.

The income statement provides detailed information on a company's revenue, expenses, and profits for a specific period. By examining this statement, investors can determine the company's financial health and identify potential areas of concern.

On the other hand, the balance sheet offers a snapshot of a corporation's assets, liabilities, and shareholders' equity as of a particular date. By understanding a company's assets and liabilities, investors can gauge its financial stability and assess its ability to meet its financial obligations.

Lastly, the cash flow statement summarizes the impact that adjustments to the balance sheet and income statements have had on cash and other liquid assets. By examining this statement, investors can determine how a company generates and uses its cash flow, which can provide valuable insights into its financial management practices.

A thorough analysis of a company's financial statements is essential for making informed investment decisions. By understanding the information in these statements, investors can make confident recommendations and minimize their financial risks.

Growth in Both Earnings and Revenue

Examining a firm's revenue growth and earnings growth is one of the quickest methods to get a sense of how healthy the company is. Generally speaking, steady growth in both areas is a significant indicator of effective management practices within a company and a healthy business strategy. However, to have a comprehensive view of the situation, it is necessary to contrast these figures with those obtained by other businesses operating in the same sector.

Capitalization of the Market

The total worth of a firm's stock shares is referred to as the company's market capitalization or market cap. To determine it, multiply the current price of the company's stock by the total number of currently circulating shares. A firm's market capitalization can provide you with an approximate estimate of its size and an indication of how risky or volatile its stock may be.

Metrics Used in Valuation

When determining the value of a company's stock, it is essential to use valuation ratios such as price-to-earnings (P/E), price-to-sales (P/S), and price-to-book (P/B). A high price-to-earnings ratio may indicate that the company is overvalued or that the market anticipates high growth rates in the future.

Both of these interpretations are possible. It is crucial to have a solid understanding of these ratios about the larger market and the general tendencies of the business.

Trends in the Various Industries and Sectors
When gaining meaningful insights, a solid understanding of the specific industry and the larger market in which a firm operates is essential. Are there reasons to be optimistic about the industry's future? Is there a perception that the industry is "hot," or is it now unpopular? A successful investment decision can be made or lost based on the responses to these questions.

Indicators of the Macroeconomic Environment
The stock market's performance can be influenced by several macroeconomic variables, including interest rates, inflation, and GDP growth. For instance, a rise in interest rates tends to have a depressing effect on stock prices, particularly for businesses with significant debt. If you understand the larger economic background, you will be better able to foresee changes in the market.

Trends and Cycles in the Market
The stock market does not remain unchanged; it moves in cyclical patterns and follows trends. Understanding all three types of market conditions—bull markets, bear markets, and sideways markets—is fundamental. Understanding these tendencies can help investors time their investments more effectively.

Evaluation of Dangers
Conduct a thorough assessment of the dangers associated with the investment. This could include risks specific to the company, such as outstanding lawsuits, or market risks, such as geopolitical upheaval. Returns that have been adjusted for risk can offer a more accurate projection of future performance, which in turn makes it possible to make more educated choices.

The Importance of Having a Diverse Portfolio
Investing all your money in one stock or industry is risky. Investors should diversify their holdings across different asset classes or sectors to safeguard against unsystematic risks specific to a particular industry or company. Diversification is a crucial aspect of any effective investment strategy.

Interpretation of Feelings
In addition to numbers and financial ratios, market sentiment is a significant factor in the pricing of stocks. Investors can get a sense of the current sentiment surrounding a particular stock or the market by using tools such as news alerts, social media tracking, and market comments.

Analyzing businesses and the market is a complex process that entails looking at more than just one or two indications at a time. It considers a wide variety of financial measurements, sector developments, and macroeconomic environment indicators. It is essential for individuals who intend to take the Series 7 exam to demonstrate a mastery of the capacity to assess firms and markets. Understanding these components can determine profitable or losing investments, even for individual investors. Knowledge of the stock market can lead to financial gain.

Chapter 14

Practice Tests and Mock Exams

Practice Test

1. Which of the following options represents a "put"?
A) The right to buy
B) The right to sell
C) The obligation to sell
D) The obligation to buy

2. What is the full form of IPO?
A) Independent Purchase Order
B) Initial Public Offering
C) International Product Order
D) Intensive Price Optimization

3. What does the term Beta mean in investing?
A) A measure of volatility
B) A type of bond
C) A Greek letter
D) An earnings ratio

4. What is meant by a margin call?
A) A conference call with investors
B) A request for more collateral on a loan
C) A dividend payment
D) A company's annual report

5. What do we mean by diversification?
A) Investing in one sector
B) Spreading investments across various asset classes
C) Short-selling stocks
D) Specializing in commodities

6. What does the term Blue Chip stock refer to?
A) A highly speculative stock
B) A stock from a well-established company
C) A penny stock
D) A foreign stock

7. What is the full form of NASDAQ?
A) National Association of State Departments of Agriculture
B) National Association of Securities Dealers Automated Quotations
C) North American Security Department Association Quorum
D) None of the above

8. Which of the following is a fixed-income security?
A) Stock
B) Bond
C) Mutual Fund
D) ETF

9. What is a 401(k) plan?
A) A type of IRA
B) A retirement savings plan
C) An educational savings plan
D) A health savings account

10. What is the primary market?
A) Where IPOs happen
B) Where used goods are sold
C) The Forex market
D) The bond market

11. Which of the following is the meaning of EPS?
A) Earnings Per Stock
B) Earnings Per Share
C) Economic Profit Sharing
D) Effective Public Sector

12. What do futures contracts imply?
A) Agreements to exchange a set number of shares at a future date
B) Agreements to buy or sell an asset at a future date
C) Contracts for employment
D) Options contracts

13. What defines a bear market?
A) A market with rising stock prices
B) A market with falling stock prices
C) A market with stagnant stock prices
D) A market for trading bear pelts

14. What is meant by a dividend?
A) A payment to shareholders
B) A type of bond
C) A tax liability
D) A penalty for early withdrawal from a bank account

15. What exactly is the S&P 500?
A) A mutual fund
B) An index of 500 large U.S. companies
C) A type of government bond
D) A European stock exchange

16. What does insider trading mean?
A) Trading based on public information
B) Trading based on confidential, non-public information
C) Trading between friends
D) Trading within a family

17. What is a hedge fund?
A) A fund that invests in hedges
B) A type of mutual fund
C) A private investment fund
D) A government-sponsored fund

18. How is liquidity defined?
A) The ease with which an asset can be converted into cash
B) The profitability of a company
C) The debt level of a company
D) The size of a company

19. What is the meaning of a bull market?
A) A market with rising stock prices
B) A market with falling stock prices
C) A market for trading bullion
D) A market for trading cattle

20. What do you understand by a callable bond?
A) A bond that can be converted into stock
B) A bond that can be repurchased by the issuer before it matures
C) A bond that pays a variable interest
D) A bond issued by the government}

21. Which of the following is an ETF?
A) Electronic Transfer Fund
B) Exchange-Traded Fund
C) Extra Tax Fee
D) External Trust Fund

22. In the context of bonds, what does yield mean?
A) Price of the bond
B) Coupon rate
C) Profitability
D) Annual interest as a percentage of the bond's current price

23. What is short selling?
A) Selling an asset you own
B) Borrowing an asset to sell, hoping to buy it back at a lower price
C) Selling an asset at a profit
D) Selling an asset in less than one year

24. What is a mutual fund?
A) A pool of money from individual investors
B) A type of savings account
C) A retirement fund
D) A type of corporate bond

25. What is the Dodd-Frank Act?
A) A healthcare act
B) A corporate tax law
C) A financial regulation law
D) An education act

26. What does ROI stand for?
A) Return On Investment
B) Risk Of Insolvency
C) Rate Of Interest
D) Roll Over IRA

27. What is market capitalization?
A) The total value of a company's outstanding shares of stock
B) The total debt of a company
C) The total assets of a company
D) The total revenue of a company

28. What is a REIT?
A) Real Estate Investment Trust
B) Retail Expense Investment Token
C) Risk Evaluation and Investment Tool
D) Real Earnings In Training

29. What is quantitative easing?
A) Lowering interest rates
B) Increasing interest rates
C) The central bank buying securities to increase money supply
D) The central bank selling securities to decrease money supply

30. What is the main role of the Federal Reserve?
A) To regulate stock exchanges
B) To enforce tax laws
C) To control the money supply and interest rates
D) To govern foreign trade

31. What is a blue-chip stock?
A) A highly volatile stock
B) A stock of a large, well-established company
C) A stock of a new startup
D) A stock that pays no dividends

32. What is a convertible bond?
A) A bond that can be converted into stock of the issuing company
B) A bond that can be repurchased by the issuer
C) A bond that pays a fixed interest rate
D) A bond issued by a foreign government

33. What is a margin call?
A) A request for more collateral on a margin account
B) A call to inform of stock splits
C) A call to offer new investment options
D) A request for early retirement

34. What is diversification?
A) Investing all in one asset class
B) Spreading investment risks by buying different types of assets
C) Focusing on short-term investments
D) Focusing on bonds only

35. What does NASDAQ stand for?
A) National Association of Securities Dealers Automated Quotations
B) North American Securities and Derivatives Association
C) New Age Securities Derivatives and Queries
D) National Automated Stock Dealers and Queries

36. What is a 401(k)?
A) A type of stock
B) A retirement savings plan sponsored by an employer
C) A government bond
D) A tax form

37. What does IPO stand for?
A) Internal Product Offering
B) Initial Public Offering
C) Investment Profit Option
D) Initial Price Option

38. What is a bid price?
A) The price a buyer is willing to pay for a security
B) The price a seller is asking for a security
C) The market price of a security
D) The final transaction price of a security

39. What is a stop-loss order?
A) An order to buy a stock when it reaches a certain price
B) An order to sell a stock when it reaches a certain price
C) An order to stop all trading activities
D) An order to buy more stocks at market price

40. What is a portfolio?
A) A collection of personal belongings
B) A collection of financial investments like stocks, bonds, and cash
C) A collection of real estate properties

41. What exactly is an option?
A) A bond category
B) A privilege to buy or sell an asset at an agreed-upon price before a specified date
C) A mutual fund type
D) A fixed rate of interest for loans

42. What is the primary market?
A) Where investors purchase securities directly from the company
B) Where investors swap securities amongst themselves
C) A commodities market variety
D) A market for derivatives

43. What is meant by the term bear market?
A) A market in which asset prices are escalating
B) A market in which asset prices are plummeting
C) A market that specializes in trading animal-related stocks
D) A market that has unpredictable price movements

44. What is a dividend?
A) An annual charge for holding a stock
B) A portion of profits paid to shareholders
C) The initial cost of a stock
D) A stock option type

45. What is a closed-end fund?
A) A fund that is not available to new investors
B) A fund that releases a fixed number of shares
C) A fund that tracks an index
D) A fund that invests in start-ups

46. What is a callable bond?
A) A bond that cannot be redeemed before maturity
B) A bond that can be converted into equity
C) A bond that can be redeemed by the issuer before maturity
D) A bond that pays variable interest

47. What does EBITDA mean?
A) Earnings Before Interest, Taxes, Depreciation, and Amortization
B) Earnings By Internal Tax Department Assessment
C) Earnings Before Investment, Trade, Dividends, and Assets
D) Estimated Business Income, Taxes, Duties, and Assessment

48. What is a hedge fund?
A) A mutual fund type open to retail investors
B) An investment fund that utilizes different strategies to earn active returns
C) A fund that only invests in agricultural commodities
D) A government-managed retirement fund

49. What is the P/E ratio?
A) Profit to Earnings ratio
B) Price to Earnings ratio
C) Price to Equity ratio
D) Profit to Equity ratio}

50. What is insider trading?
A) Trading based on public information
B) Trading based on confidential, non-public information
C) Purchasing a large amount of shares at once
D) Selling a large amount of shares at once

51. Which of the following describes a junk bond?
A) A bond that has defaulted
B) A bond rated below investment grade
C) A bond that offers high interest
D) A bond issued by a tech startup

52. What is short selling?
A) Selling shares you currently own
B) Buying shares with borrowed money
C) Selling borrowed shares with the intention of buying them back at a lower price
D) Buying and selling shares on the same day

53. What is ETF short for?
A) Estimated Time of Funding
B) Exchange-Traded Fund
C) Exponential Trading Factor
D) Earnings Through Franchise

54. What is day trading?
A) Trading shares multiple times within a single trading day
B) Holding shares for more than a year
C) Holding shares for less than a month
D) Trading shares only on Mondays

55. What is an accredited investor?
A) An investor with special regulatory approval
B) An investor who has passed the Series 7 exam
C) An investor who meets specific income or net worth requirements
D) An investor with a business degree

56. What is an annuity?
A) A lump-sum payment
B) A series of payments made at regular intervals
C) A type of life insurance
D) A type of loan

57. What is a mutual fund?
A) A fund that trades on an exchange like a stock
B) A pool of funds from many investors that is managed by an investment company
C) A fund that only invests in government securities
D) A fund that only contains blue-chip stocks

58. What does ROI stand for?
A) Rate of Inflation
B) Rate of Interest
C) Return on Investment
D) Roll Over IRA

59. What is a market order?
A) An order to buy or sell at a specific price
B) An order to buy or sell immediately at the current market price
C) An order that only executes at market close
D) An order that is valid until canceled

60. What is an illiquid asset?
A) An asset that is easy to buy and sell
B) An asset that is difficult to buy and sell
C) An asset that provides high returns
D) An asset that is prone to default

61. Which of the following best defines bid price?
A) The price at which an investor is willing to sell a security
B) The highest price a security has reached in a trading day
C) The price at which an investor is willing to buy a security
D) The lowest price a security has reached in a trading day

62. What is the meaning of blue-chip stock?
A) A stock from a small, growth-oriented company
B) A stock from a large, stable, and financially sound company
C) A stock that pays a high dividend
D) A stock in the technology sector

63. What is the full form of NASDAQ?
A) National Association of Securities Dealers Automated Quotations
B) North American Securities and Derivatives Association
C) National Assets and Securities Technologies Association Quotations
D) North American Stock Dealers Association Quotations

64. What is a margin call?
A) A request for additional funds when a margin account falls below the minimum required
B) A call to buy more shares of a specific stock
C) A call to sell all positions in a margin account
D) A call to confirm a trade

65. What is a put option?
A) The right to buy a stock at a specified price before a certain date
B) The right to sell a stock at a specified price before a certain date
C) An option that increases in value as the stock price rises
D) An option that expires immediately if the stock price falls

66. What is the SIPC?
A) A government agency that insures investment accounts
B) A private corporation that insures investors' brokerage accounts
C) A government agency that regulates the stock market
D) A private corporation that regulates investment advisors

67. What is diversification?
A) Investing all money into one asset class
B) Investing in different types of assets to reduce risk
C) Investing in only blue-chip stocks
D) Investing in one sector of the economy

68. What is a stop order?
A) An order that turns into a market order once a certain price is reached
B) An order to buy or sell at a specific price
C) An order to sell all assets immediately
D) An order to stop trading a specific stock

69. What is market capitalization?
A) The total dollar market value of a company's shares
B) The total assets of a company
C) The capital raised through an IPO
D) The total liabilities of a company

70. What is a convertible bond?
A) A bond that can be exchanged for a specific number of shares of the company's stock
B) A bond that can be redeemed by the issuer before its maturity date
C) A bond that has defaulted
D) A bond that pays a variable rate of interest

71. What is a bear market?
A) Stock prices are rising.
B) Stock prices are falling.
C) There is high volatility.
D) There is lots of investor activity.

72. What is the meaning of IPO?
A) Initial Price Offering
B) Immediate Public Offering
C) Initial Public Offering
D) Internal Process Order

73. What is a call option?
A) The right to sell a stock at a specified price before a certain date.
B) The right to buy a stock at a specified price before a certain date.
C) An option that expires immediately if the stock price rises.
D) The obligation to buy a stock at a specified price before a certain date.

74. What is a bond's coupon rate?
A) The rate at which the bond's price fluctuates.
B) The interest rate paid on the bond.
C) The rate at which the bond is converted into stock.
D) The rate at which the bond matures.

75. What is a dividend?
A) A one-time payment made by a company to its shareholders.
B) A share of profits paid regularly by a company to its shareholders.
C) A reduction in the price of a company's stock.
D) A payment made by a company to pay off its debt.

76. What is a hedge fund?
A) A fund that only invests in conservative, low-risk assets.
B) A private investment fund that uses a variety of strategies to earn returns.
C) A fund that only invests in commodities like gold and oil.
D) A fund that only holds government bonds.

77. What is a bull market?
A) Stock prices are falling.
B) Stock prices are rising.
C) There is low investor activity.
D) There is high volatility.

78. What is liquidity?
A) The ability of an asset to generate cash quickly.
B) The stability of an asset's price.
C) The rate of return on an asset.
D) The level of risk associated with an asset.

79. What is a preferred stock?
A) A stock that has voting rights.
B) A stock that pays fixed dividends but has no voting rights.
C) A stock that pays higher dividends than common stock.
D) A stock that only institutional investors can purchase.

80. What is a capital gain?
A) The profit made from the sale of a capital asset.
B) The interest earned on a bond.
C) The dividends paid on a stock.
D) The original investment amount in a security.

81. What is the meaning of ETF?
A) Estimated Tax Fund
B) Electronic Trade Financing
C) Exchange-Traded Fund
D) Equity Transfer Facility

82. Define index.
A) A type of mutual fund
B) A benchmark for measuring the performance of an asset class
C) A type of bond
D) A trading platform

83. What is a market order?
A) An order to buy or sell a security at a specific price
B) An order to buy or sell a security immediately at the current market price
C) An order to buy a security once it reaches a certain price
D) An order to sell a security once it falls to a certain price

84. What is a mutual fund?
A) An investment vehicle that holds a diversified portfolio and is managed by a professional
B) A single stock or bond
C) A type of bank account
D) A trading strategy

85. Explain P/E ratio.
A) Price to Equity
B) Profit to Earnings
C) Price to Earnings
D) Portfolio to Equity

86. What is alpha?
A) The average return on an investment
B) The excess return of an investment compared to its benchmark
C) The risk associated with an investment
D) The dividend yield of a stock

87. What does REIT mean?
A) Real Estate Investment Trust
B) Renewable Energy Income Trust
C) Retail Earnings Income Tax
D) Rate of Earnings and Inflation Target

88. Define growth stock.
A) A stock that pays a high dividend
B) A stock that has a history of stable growth
C) A stock from a company expected to grow at an above-average rate
D) A stock that has a high P/E ratio

89. What is a day trader?
A) An investor who holds positions for the long term
B) A trader who buys and sells securities within the same trading day
C) A trader who focuses only on options
D) A trader who only trades once a week

90. What does SEC stand for?
A) Securities Exchange Corporation
B) Security and Equity Council
C) Stock Exchange Committee
D) Securities and Exchange Commission

91. Define insider trading.
A) Trading based on public information
B) Trading by company executives based on non-public information
C) Trading by day traders
D) Trading based on analyst recommendations

92. Define leverage.
A) The use of borrowed money to increase investment returns
B) The process of selling securities short
C) The cost of trading a security
D) The percentage of a portfolio in stocks

93. Define blue-chip stock.
A) A stock from a small-cap company
B) A stock from a well-established and financially sound company
C) A stock with a high dividend yield
D) A newly listed stock

94. Define the put option.
A) The right to buy a stock at a certain price
B) The right to sell a stock at a certain price
C) An obligation to buy a stock at a certain price
D) An obligation to sell a stock at a certain price

95. What is the meaning of ROI?
A) Return on Investment
B) Rate of Inflation
C) Return on Interest
D) Rate of Interest

96. What is the bid price?
A) The price a buyer is willing to pay for a security
B) The price a seller is willing to accept for a security
C) The highest price a security has reached
D) The lowest price a security has reached

97. Define diversification.
A) Investing all funds in a single asset
B) Investing in a range of different assets to reduce risk
C) Investing only in high-risk assets
D) Investing only in low-risk assets

98. What is a margin account?
A) An account used only for trading options
B) An account that allows you to borrow money for investing
C) An account used only for trading stocks
D) An account for retirement savings

99. What is capital loss?
A) The loss incurred when an investment decreases in value
B) The loss incurred when an investment increases in value
C) The cost of trading a security
D) The interest paid on a loan

100. Define limit order.
A) An order to buy or sell a security at the current market price
B) An order to buy or sell a security at a specific price or better
C) An order that becomes a market order once a certain price is reached
D) An order that is only valid for the day

101. What is the main purpose of an IPO (Initial Public Offering)?
A) To enable a company to repurchase its shares
B) To enable a company to become public and raise capital
C) To enable a company to liquidate its assets
D) To enable a company to pay dividends

102. What is the intrinsic value of an option?
A) The time left until it expires
B) The option's premium
C) The difference between the option's strike price and the underlying asset's market price
D) The option's volatility

103. What does the term "Blue Chip Stock" mean?
A) A stock that has recently become public
B) A stock from a well-established and financially stable company
C) A stock with high growth potential
D) A stock with high dividends

104. What is a callable bond?
A) A bond that can be converted into stock
B) A bond that can be redeemed by the issuer before its maturity
C) A bond that pays coupons monthly
D) A bond with no fixed maturity date

105. What does the P/E ratio measure?
A) Profit per share
B) Earnings volatility
C) Market capitalization
D) Price relative to earnings

106. What is a "limit order"?
A) An order to buy or sell a security instantly at the current price
B) An order to buy or sell a security at a specific price or better
C) An order to sell a security if it reaches a certain price
D) An order to buy a security with borrowed funds

107. What is a "bid-ask spread"?
A) The difference between the highest price a buyer is willing to pay and the lowest price a seller is willing to accept
B) The range of prices in an auction
C) The cost of an option contract
D) The difference between opening and closing prices

108. What does "stop-loss order" mean?
A) An order to buy a security at a higher price
B) An order to sell a security if it falls to a certain price
C) An order to buy a security if it reaches a certain price
D) An order to hold a security indefinitely

109. What is "dollar-cost averaging"?
A) Investing a fixed dollar amount at regular intervals, regardless of share price
B) Investing only when the market is in a downtrend
C) Buying stocks in increments of $100
D) Selling stocks when they have appreciated by a dollar amount

110. What is "liquidity"?
A) The ease with which an asset can be converted into cash
B) The profitability of an investment
C) The stability of a company's earnings
D) The safety of an investment

111. What is the meaning of "short selling"?
A) Expecting a security's price to increase by purchasing it
B) Selling a security that is not owned
C) Intending to sell a security quickly after purchasing it
D) Selling a security with a stop-loss order

112. What defines a "zero-coupon bond"?
A) A bond that does not pay interest
B) A bond that pays interest only at maturity
C) A bond that pays interest annually
D) A bond that pays interest monthly

113. What is a dividend?
A) The capital gain of an asset
B) The interest earned from a loan
C) A distribution of a portion of a company's earnings to shareholders
D) The expense ratio of an ETF

114. What is the debt-to-equity ratio?
A) A measure of a company's profitability
B) A measure of a company's financial leverage
C) A measure of a company's liquidity
D) A measure of a company's market capitalization

115. What is meant by "portfolio rebalancing"?
A) The process of purchasing more of the top-performing assets in your portfolio
B) The process of buying and selling assets to maintain your desired asset allocation
C) The process of converting all your assets into cash
D) The process of moving all your investments into a single asset class

116. What is the definition of "market capitalization"?
A) The total value of a company's outstanding shares of stock
B) The total value of a company's assets
C) The total value of a company's annual earnings
D) The total value of a company's debt

117. What is a "hedge fund"?
A) A fund that invests only in commodities
B) A fund that invests only in bonds
C) A private investment fund that employs various strategies to earn returns
D) A fund that invests only in blue-chip stocks

118. What is "front-running"?
A) Buying or selling securities before a large client trade is executed
B) Buying or selling securities after market hours
C) Using leverage to amplify returns
D) Trading based on insider information

119. What is the definition of "spread" in options trading?
A) The difference between the bid and ask prices
B) The range of possible strike prices
C) Buying and selling options simultaneously
D) The time decay of an option's price

120. What is meant by "vesting" in the context of employee benefits?
A) The immediate availability of benefits
B) The process by which an employee gains access to benefits over time
C) The termination of benefits after employment
D) The inflation adjustment for retirement benefits

121. What is the definition of "arbitrage"?
A) Purchasing a security in one market and selling it in another in order to make a profit from a price difference
B) The act of short selling a stock
C) Buying a stock based on insider information
D) Using derivatives to manage risk

122. What is meant by "underwriting"?
A) The process of evaluating the risk of insuring a specific person or asset
B) The act of guaranteeing the sale of a new issue of securities
C) The process of rebalancing a portfolio
D) The act of transferring securities from one account to another

123. What is a "mutual fund"?
A) A fund that is managed by a single investor
B) A pooled investment vehicle managed by an investment company
C) A type of insurance policy that covers multiple people
D) A fund that only invests in bonds

124. What is the meaning of the term "leverage" in the context of finance?
A) The use of borrowed capital to increase the potential return of an investment
B) The use of technical analysis to predict stock movements
C) The use of a balanced portfolio to mitigate risk
D) The use of derivative products to speculate on market trends

125. What is a "margin call"?
A) A request from a broker to deposit more money or securities into a margin account
B) A phone call from a broker pitching a new investment opportunity
C) A fee charged when opening a margin account
D) A call option bought on margin

CHAPTER 15

ANSWER KEYS AND EXPLANATIONS

1. The correct answer is B, which stands for the right to sell. A put option gives the holder the right, but not the obligation, to sell an underlying asset at a specific price within a specific timeframe.

2. The correct answer is B, which stands for Initial Public Offering (IPO). This is when a private company publishes shares to the general public for the first time.

3. The correct answer is A, a measure of volatility. Beta measures a stock's volatility about the overall market.

4. The correct answer is B, a request for more collateral on a loan. A margin call occurs when an investor must deposit more money into a margin account after a decline in the value of the assets being held as collateral.

5. The correct answer is B, which means spreading investments across various asset classes. Diversification is spreading investments across various asset classes to minimize risk.

6. The correct answer is B, which stands for a stock from a well-established company. A Blue Chip stock typically comes from a large, well-established company with a history of reliable performance.

7. The correct answer is B, which stands for National Association of Securities Dealers Automated Quotations (NASDAQ). It is an American stock exchange.

8. The correct answer is B, which stands for bond. Bonds are fixed-income securities that pay a fixed interest over a certain period.

9. The correct answer is B, which stands for a retirement savings plan. A 401(k) is a retirement savings plan sponsored by an employer.

10. The correct answer is A, which is where IPOs happen. The primary market is where new securities are issued and sold to the public, often through IPOs.

11. The correct answer is B, EPS is calculated by dividing a company's profit by the number of outstanding shares of its common stock.

12. The correct answer is Agreements to buy or sell an asset at a future date, which means futures contracts are standardized agreements to buy or sell a specific asset at a predetermined price at a specified time in the future.

13. The correct answer is A market with falling stock prices, which means a bear market is a market characterized by falling stock prices, usually by 20% or more from recent highs.

14. The correct answer is A payment to shareholders, which means a dividend is a payment made by a corporation to its shareholders, usually in the form of cash or additional stock.

15. The correct answer is An index of 500 large U.S. companies, which means the S&P 500 is a market-capitalization-weighted index of 500 large U.S. companies considered to be leading indicators of the U.S. stock market.

16. The correct answer is Trading based on confidential, non-public information, which means insider trading refers to the illegal practice of trading stocks or other securities based on material, non-public information.

17. The correct answer is A private investment fund, which means a hedge fund is a private investment fund that employs various strategies to generate returns for its investors.

18. The correct answer is The ease with which an asset can be converted into cash, which means liquidity refers to the ease with which an asset or security can be quickly bought or sold without affecting its price.

19. The correct answer is A market with rising stock prices, which means a bull market is a market characterized by rising stock prices, usually for an extended period of time.

20. The correct answer is A bond that can be repurchased by the issuer before it matures, which means a callable bond is a bond that can be redeemed or "called" by the issuer before its maturity date, usually at a specified call price.

21. The correct answer is Exchange-Traded Fund, which means a type of investment fund that can be traded on stock exchanges, similar to individual stocks.

22. The correct answer is A, an annual interest as a percentage of the bond's current price, which means yield refers to the annual return on an investment, expressed as a percentage based on the investment's cost, current market value, or face value. In the case of bonds, it's the annual interest paid as a percentage of the bond's current market price.

23. The correct answer is Borrowing an asset to sell, hoping to buy it back at a lower price, which means short selling involves borrowing an asset, selling it, and then buying it back later at a lower price to return to the lender, profiting from the price difference.

24. The correct answer is A pool of money from individual investors, which means a mutual fund pools money from various investors to buy securities such as stocks, bonds, and other assets.

25. The correct answer is A financial regulation law, which means the Dodd-Frank Act is a comprehensive financial reform law passed in 2010 in response to the financial crisis of 2008.

26. The correct answer is Return On Investment, which means ROI stands for Return On Investment, a measure used to evaluate the profitability of an investment.

27. The correct answer is The total value of a company's outstanding shares of stock, which means market capitalization is calculated by multiplying a company's stock price by its number of outstanding shares, providing a total value for the company's equity.

28. The correct answer is Real Estate Investment Trust, which means a REIT (Real Estate Investment Trust) is a company that owns, operates, or finances income-producing real estate across a range of property sectors.

29. The correct answer is The central bank buying securities to increase money supply, which means quantitative easing is a non-traditional monetary policy where a central bank purchases financial assets to increase the money supply and lower interest rates.

30. The correct answer is To control the money supply and interest rates, which means the primary role of the Federal Reserve is to regulate the U.S. monetary and financial system, including controlling the money supply and setting interest rates to achieve economic objectives like stable prices and full employment.

31. The correct answer is A stock of a large, well-established company, which means blue-chip stocks are shares in large, well-established companies with a history of stable, reliable performance.

32. The correct answer is: A bond that can be converted into stock of the issuing company, which means a convertible bond is a type of bond that gives the bondholder the right to convert the bond into shares of the issuing company.

33. The correct answer is A request for more collateral on a margin account, which means a margin call is a broker's demand for an investor to deposit additional money or securities to cover potential losses.

34. The correct answer is Spreading investment risks by buying different types of assets, which means diversification involves spreading your investments across different types of assets to reduce risk.

35. The correct answer is National Association of Securities Dealers Automated Quotations, which means NASDAQ stands for National Association of Securities Dealers Automated Quotations. It's an American stock exchange.

36. The correct answer is A retirement savings plan sponsored by an employer, which means a 401(k) is a retirement savings plan sponsored by an employer. It allows workers to save and invest a piece of their paycheck before taxes are taken out.

37. The correct answer is Initial Public Offering, which means IPO stands for Initial Public Offering. It's the first sale of stock by a private company to the public.

38. The correct answer is The price a buyer is willing to pay for a security, which means the bid price is the highest price that a prospective buyer is willing to pay for a specific security.

39. The correct answer is An order to sell a stock when it reaches a certain price, which means a stop-loss order is an order placed to sell a given stock once it reaches a certain price to mitigate losses.

40. The correct answer is A collection of financial investments like stocks, bonds, and cash, which means a portfolio collects financial investments, including stocks, bonds, and cash equivalents.

41. The correct answer is: An option is a financial instrument that gives the holder the right, but not the obligation, to buy or sell an asset at a specific price on or before a certain date, which means an option grants the right but not the obligation to execute a transaction under specified conditions.

42. The correct answer is The primary market is the part of the capital market that deals with the issuance of new securities, which means investors buy securities directly from the company issuing them.

43. The correct answer is: A bear market is characterized by falling prices for a prolonged period of time, usually by 20% or more from recent highs, which means a bear market signifies a prolonged downturn in a financial market.

44. The correct answer is: A dividend is a distribution of a portion of a company's earnings to its shareholders, which means dividends can be in the form of cash or additional shares.

45. The correct answer is A closed-end fund is a publicly traded investment company that raises a fixed amount of capital, which means it issues a fixed number of shares via an IPO.

46. The correct answer is A callable bond is a bond that the issuer has the right to redeem before its maturity date, which means the issuer can buy back the bond under certain conditions.

47. The correct answer is EBITDA stands for Earnings Before Interest, Taxes, Depreciation, and Amortization, which means it's a measure often used to evaluate a company's operational performance.

48. The correct answer is A hedge fund is an investment fund that employs various strategies, which means it trades in relatively liquid assets and can use more complex trading and risk management techniques.

49. The correct answer is The P/E ratio (Price to Earnings ratio) is a valuation ratio, which means it's calculated by dividing the market value per share by the earnings per share.

50. The correct answer is Insider trading refers to the practice of trading a public company's stock based on material, non-public information, which means this activity is typically illegal and unethical.

51. The correct answer is A junk bond is a bond rated below investment grade, which means they offer higher yields to compensate for their higher risk.

52. The correct answer is Short selling involves selling borrowed shares with the intent to buy them back at a lower price, which means the goal is to profit from falling prices.

53. The correct answer is ETF stands for Exchange-Traded Fund, which means it's a type of investment fund that is traded on a stock exchange.

54. The correct answer is Day trading is the practice of buying and selling financial instruments within the same trading day, which means the objective is to profit from short-term price movements.

55. The correct answer is: An accredited investor is an individual or entity that meets certain income or net worth criteria, which means they can participate in higher-risk investments.

56. The correct answer is An annuity is a series of payments made at regular intervals, which means it's often used for retirement income.

57. The correct answer is A mutual fund pools money from many investors to purchase a diversified portfolio of assets, which means it's managed by an investment company.

58. The correct answer is ROI stands for Return on Investment, which means it's a measure used to evaluate the profitability of an investment.

59. The correct answer is A market order is an order to buy or sell a security immediately at the current market price, which means it doesn't specify a price limit.

60. The correct answer is An illiquid asset is one that cannot be easily bought or sold, which means selling it could substantially affect its price.

61. The correct answer is: The bid price is the price at which an investor is willing to buy a security, which means it's the highest price that a buyer is willing to pay for an asset.

62. The correct answer is: A blue-chip stock is a stock from a large, stable, and financially sound company, which means it has a history of providing reliable returns.

63. The correct answer is NASDAQ stands for National Association of Securities Dealers Automated Quotations, which means it's an electronic marketplace where investors can buy and sell stocks.

64. The correct answer is A margin call is a request for additional funds, which means it's made when a margin account falls below the minimum required.

65. The correct answer is: A put option is the right to sell a stock at a specified price before a certain date, which means it gives the holder the option but not the obligation to sell a stock.

66. The correct answer is The SIPC is a private corporation that insures investors' brokerage accounts, which means it provides protection against the failure of the brokerage firm up to certain limits.

67. The correct answer is Diversification is investing in different types of assets, which means it's a strategy used to reduce risk.

68. The correct answer is: A stop order is an order that turns into a market order once a certain price is reached, which means it becomes executable when a specific price condition is met.

69. The correct answer is Market capitalization is the total dollar market value of a company's shares, which means it represents the company's overall value in the stock market.

70. The correct answer is: A convertible bond is a bond that can be exchanged for a specific number of shares, which means it gives the holder the option to convert the bond into stock.

71. The correct answer is a bear market is a condition where stock prices are falling, generally by 20% or more from recent highs, which leads to widespread pessimism.

72. The correct answer is IPO stands for Initial Public Offering, indicating the first time a company's shares are offered to the public, transitioning it from a private to a public entity.

73. The correct answer is a call option gives you the right, but not the obligation, to buy an asset at a predetermined price within a specified timeframe.

74. The correct answer is the coupon rate is the annual interest rate paid to bondholders by the bond issuer, usually expressed as a percentage.

75. The correct answer is a dividend is a distribution of a portion of a company's earnings to shareholders, usually in the form of cash or additional shares.

76. The correct answer is a hedge fund is a pooled investment fund that employs various strategies to earn returns for its investors.

77. The correct answer is a bull market is a market condition where there is a prolonged period of rising asset prices, usually by 20% or more from recent lows.

78. The correct answer is liquidity refers to the ability to quickly convert an asset into cash without a significant impact on its market price.

79. The correct answer is preferred stock is a type of stock that has a higher claim on assets and earnings than common stock and usually does not have voting rights.

80. The correct answer is a capital gain is the profit earned when you sell an asset for more than you paid for it.

81. The correct answer is ETF stands for Exchange-Traded Fund, a type of investment fund that can be bought and sold on stock exchanges like individual stocks.

82. The correct answer is an index is a benchmark used to measure and compare the performance of various assets or asset classes.

83. The correct answer is a market order is an instruction to buy or sell a security as soon as possible at the current market price.

84. The correct answer is a mutual fund is a type of investment vehicle that pools funds from various investors to purchase a diversified portfolio of stocks, bonds, or other securities.

85. The correct answer is P/E ratio stands for Price to Earnings Ratio, which measures a company's current share price relative to its earnings per share.

86. The correct answer is alpha measures the performance of an investment against a market index or benchmark; it indicates how much an investment has outperformed or underperformed its market index.

87. The correct answer is REIT stands for Real Estate Investment Trust, a company that owns, finances, or operates income-producing real estate.

88. The correct answer is a growth stock is a share in a company that is expected to experience above-average growth compared to other stocks in the market.

89. The correct answer is a day trader is someone who buys and sells financial instruments within the same trading day, aiming to profit from short-term price movements.

90. The correct answer is SEC stands for Securities and Exchange Commission, the U.S. government agency responsible for regulating the securities industry and protecting investors.

91. The correct answer is insider trading involves the illegal practice of trading a public company's stock or other securities based on material, non-public information about the company.

92. The correct answer is leverage is the use of borrowed money or financial instruments to amplify the potential return of an investment, but it also increases the risk associated.

93. The correct answer is a blue-chip stock refers to shares in a large, reputable company known for its reliability, quality, and ability to operate profitably.

94. The correct answer is a put option grants the owner the right, but not the obligation, to sell a specific amount of an underlying asset at a set price within a specific time frame.

95. The correct answer is ROI stands for Return on Investment, a metric used to evaluate the financial returns of an investment relative to its cost.

96. The correct answer is the bid price is the highest price a buyer is willing to pay for a security.

97. The correct answer is diversification is the investment strategy of spreading money among different assets or asset classes to reduce risk.

98. The correct answer is a margin account allows an investor to purchase securities with funds borrowed from a broker, with the account acting as collateral.

99. The correct answer is a capital loss occurs when you sell an asset for less than what you initially paid for it.

100. The correct answer is a limit order is an instruction to buy or sell a security at a specified price or better, unlike a market order, which trades at the current price.

101. The correct answer is the purpose of an IPO is to enable a company to go public and raise capital by offering its shares for sale to the general public.

102. The correct answer is the intrinsic value of an option is the difference between the option's strike price and the current market price of the underlying asset when the option is in-the-money.

103. The correct answer is Blue Chip Stocks are equities from well-established and financially stable companies that have a history of yielding good returns to investors.

104. The correct answer is a callable bond is a bond that the issuer has the right to redeem before its maturity date, generally at a price above its face value.

105. The correct answer is the P/E ratio measures the market value of a stock in relation to its earnings, and is often used for valuation purposes.

106. The correct answer is a limit order is an order to buy or sell a security at a specific price or better, and will only be executed if the market reaches that price.

107. The correct answer is the bid-ask spread is the gap between the highest price a buyer is willing to pay and the lowest price a seller is willing to accept.

108. The correct answer is a stop-loss order is an order that gets triggered to sell a security when it reaches a specific price, in an attempt to limit potential losses.

109. The correct answer is dollar-cost averaging is a strategy where a fixed amount of money is invested at regular intervals, regardless of the price of the asset.

110. The correct answer is liquidity refers to the capability of an asset to be quickly converted into cash without causing a significant impact on its market price.

111. The correct answer is short selling involves borrowing shares of a stock and selling them, hoping to buy them back later at a lower price to make a profit.

112. The correct answer is a zero-coupon bond is a type of bond that doesn't pay interest but is sold at a discount to its face value. It matures at its face value.

113. The correct answer is dividends are the payouts that companies give to their shareholders, generally either in cash or in the form of more shares.

114. The correct answer is the debt-to-equity ratio measures how a company is financing its operations through debt versus equity, providing insight into its financial leverage.

115. The correct answer is portfolio rebalancing is the act of re-adjusting the allocation of assets in a portfolio to maintain the original or desired level of risk and asset balance.

116. The correct answer is market capitalization, or market cap, is the total market value of a company's outstanding shares, calculated by multiplying the share price by the number of outstanding shares.

117. The correct answer is hedge funds are investment funds that are less regulated than traditional funds, and often employ riskier strategies to earn higher returns.

118. The correct answer is front-running is the unethical practice where a broker trades an asset based on insider knowledge of future transactions that will influence its price, usually to the detriment of their clients.

119. The correct answer is an options spread is a trading strategy in which you buy and sell different options contracts simultaneously, which can involve calls, puts, or a mix of both.

120. The correct answer is vesting is the period over which an employee gains full access to benefits or stock options provided by their employer.

121: The correct answer is Arbitrage involves buying a security in one market and selling it in another to take advantage of differing prices, aiming to make a profit.

122: The correct answer is Underwriting is the guarantee by an investment bank to sell a company's new issue of securities, providing liquidity and raising capital for the company.

123: The correct answer is A mutual fund is an investment vehicle that pools money from multiple investors to buy a diversified portfolio of stocks, bonds, or other assets, managed by professionals.

124: The correct answer is Leverage is the use of borrowed capital to enhance the potential returns of an investment, while also increasing the associated risks.

125: The correct answer is: A margin call is a request from a broker for an investor to deposit more funds or securities into a margin account because its value has dropped below a minimum requirement.

CONCLUSION

When it comes to finance and investing, many intricacies and complexities must be understood to succeed in the securities sector. It is no easy task to achieve this level of mastery, but it is crucial for anyone who aspires to make a name for themselves in this field. Fortunately, there is a way to prepare for this challenge and take the first step towards a fulfilling career that will test your limits and provide you with a sense of accomplishment. That first step is studying for the Series 7 exam.

The comprehensive Series 7 exam covers a broad range of topics related to the securities industry, ensuring that those who pass it thoroughly understand the various products and strategies involved in buying and selling securities. This includes stocks, bonds, mutual funds, and options. While it can be challenging and requires a lot of hard work and dedication, passing the exam is a necessary step for anyone looking to become a licensed securities professional.

To prepare for the Series 7 exam, there are several resources available. Many people enroll in a prep course or hire a tutor to help them study. Numerous study guides and practice tests are available online that can help you prepare for the exam. The key is to find a study method that works best for you and stick with it.

Ultimately, passing the Series 7 exam is the first step towards a successful career in the securities industry. Once you have your license, there are many opportunities available to you. Whether you work for a large brokerage firm or strike out on your own as an independent financial advisor, the skills and knowledge you gain from studying for the Series 7 exam will serve you well throughout your career. So, if you're serious about making a name for yourself in finance and investing, start studying for the Series 7 exam today!

This book is designed to be your go-to reference, breaking down difficult concepts into easy-to-understand language and presenting them in a question-and-answer format that aligns exactly with the actual exam. With the information in this guide, you will gain the knowledge, techniques, and, most importantly, the self-assurance you need to pass the Series 7 test.

This guide will help you comprehend the complexities of equities and bonds and delve deeply into financial derivatives and portfolio management. However, it is essential to remember that the finance business is constantly moving forward. The field is continually changing due to shifts in market conditions, technological developments, and modifications to laws and regulations.

As you prepare to take the test, it is important to remember that this is only the beginning of the material. While completing Series 7 is a significant milestone, it is just one step along a much longer road of learning and adapting throughout one's life. Even after obtaining the General Securities Representative license, it is crucial to continue learning new things, engage in conversations, and seek additional perspectives.

Your path to a successful career in the financial industry will be paved with both opportunities and challenges. However, with the knowledge and insights from reading this book, you are better equipped to navigate this path and tackle the Series 7 exam head-on. Your time spent working in finance is as rewarding as it is profitable, and you continue to learn and grow in this exciting field.

Made in the USA
Las Vegas, NV
11 June 2024